AFRICAN MOLE-RATS: ECOLOGY AND EUSOCIALITY

African mole-rats are a unique taxon of subterranean rodents that range in sociality from solitary-dwelling species through to two 'eusocial' species, the Damaraland mole-rat and the naked mole-rat. The naked mole-rat is arguably the closest that a mammal comes to behaving like social insects such as bees and termites, with large colonies and a behavioural and reproductive division of labour. As a family, the Bathyergidae represent a model system with which to study the evolution and maintenance of highly social cooperative breeding strategies. Here, Nigel Bennett and Chris Faulkes provide a synthesis of the current knowledge of bathyergid systematics, ecology, reproductive biology, behaviour and genetics. With this, they explore the role of these factors in the evolution of sociality in the Bathyergidae in the context of both vertebrates and invertebrates. This will be an important new resource for anyone interested in the evolution of sociality, and in mole-rats in particular.

NIGEL C. BENNETT is Professor of Zoology in the Department of Zoology and Entomology at the University of Pretoria, Republic of South Africa.

CHRIS G. FAULKES is Lecturer in Molecular Ecology and Evolution in the school of Biological Sciences at Queen Mary and Westfield College, University of London.

AFRICAN MOLE-RATS: ECOLOGY AND EUSOCIALITY

by

NIGEL C. BENNETT
Department of Zoology & Entomology,
University of Pretoria,
South Africa

and

CHRIS G. FAULKES
Zoological Society of London
and
School of Biological Sciences,
Queen Mary & Westfield College,
University of London,
United Kingdom

CAMBRIDGE UNIVERSITY PRESS
Cambridge, New York, Melbourne, Madrid, Cape Town, Singapore, São Paulo

Cambridge University Press
The Edinburgh Building, Cambridge CB2 2RU, UK

Published in the United States of America by Cambridge University Press, New York

www.cambridge.org
Information on this title: www.cambridge.org/9780521771993

First published 2000
This digitally printed first paperback version 2005

A catalogue record for this publication is available from the British Library

ISBN-13 978-0-521-77199-3 hardback
ISBN-10 0-521-77199-4 hardback

ISBN-13 978-0-521-01865-4 paperback
ISBN-10 0-521-01865-X paperback

For my late mother Jean Bennett, and Fernanda and the children

and

For Lorna and Pippa

Contents

Foreword

The Bathyergidae, a family of subterranean rodents endemic to Africa, have received far more attention than would appear merited by their numbers or visibility. Initially, interest was sparked by the rather bizarre appearance, unusual physiology and social organisation of the naked mole-rat, *Heterocephalus glaber*. However, over the last decade, it has become increasingly apparent that the entire family provides fascinating insights into a wide diversity of topics, some features of which are unique to the family. The range of social behaviour exhibited by the family is wide, from strictly solitary species to what are arguably the most social mammalian species. Their subterranean lifestyle imposes significant constraints on the locating of food and of mates; they also exhibit a number of interesting physiological adapations which include suppression of reproduction in most colony members and varying degrees of thermolability, the latter culminating in the naked mole-rat, whose body temperature fluctuates with ambient temperature in a reptilian fashion. The interdisciplinary approach to research on this family, combining aspects of their behaviour, ecology, phylogeny and physiology, has been important in trying to understand the range of sociality exhibited within the family and has provided a broad platform to address the evolution of mammalian sociality in general.

Nigel Bennett and Chris Faulkes are well qualified to co-author this book. They have both been engaged in research on the Bathyergidae for more than a decade. Nigel's interest in mole-rats began with his Ph.D. dissertation in which he conducted the first detailed research on three species of haired bathyergids - the solitary Cape mole-rat, *Georychus capensis*, and two social species, the common mole-rat, *Cryptomys hottentotus hottentotus*, and the Damaraland mole-rat, *C. damarensis*. His Ph.D. was done under my supervision at the University of Cape Town and we have continued to collaborate on aspects of mole-rat biology, especially on long-term field studies on the Damaraland mole-rat. Nigel is now based at the University of Pretoria where much of his research is on aspects of mole-rat physiology. Chris Faulkes also first began research on mole-rats during his Ph.D when he examined social suppression of reproduction in the naked mole-rat. Dr David Abbott supervised his research, which was done at the Institute of Zoology, London. Subsequent to this Chris joined Dr Mike Bruford's Conservation Genetics Group at the Institute and changed the main thrust of his research from reproductive physiology to genetics. He now lectures at the Queen Mary and Westfield College, London. Chris has collaborated with Nigel and myself on many aspects of mole-rat biology and has joined us on several field trips, and this team work continues to be very fruitful. Indeed one of the strong features of research on the Bathyergidae has been the broad degree of collaboration and sharing of ideas between an ever increasing number of scientists working on mole-rats.

In this book, Nigel Bennett and Chris Faulkes have attempted to synthesise several decades of research, by a diversity of scientists, on the Bathyergidae and to present it in a way that will be intelligible to a wide readership. I am confident that the book will be of interest to many people, that it will provide a valuable source of information for many years to come, and that it will stimulate further research on these fascinating mammals.

Jennifer Jarvis
Department of Zoology, University of Cape Town

Acknowledgements

A large number of people have been involved both directly and indirectly in the production of this book, and in the research work reported here. We would especially like to thank Professor Jenny Jarvis for her input, as well as her friendship and support over many years. We are also greatly indebted to Pippa Marsden for many hours of patient proof reading and correcting earlier drafts of all the chapters, and also to Steve Le Comber and Pat Manly for final proof reading. Professor Rory Putman critically reviewed the entire manuscript, vastly improving our early versions, and offered much needed moral support during the tricky road to publication. For this we pass on a big thank you. Many others also commented and provided constructive criticism, including Andrew Bourke, Mike Bruford, Frank Clarke, Rosie Cooney, Jenny Jarvis, Steve Le Comber, Andrew Molteno, Tim Roper and Andrew Spinks. Others spent time in the field with us trapping mole-rats and collecting data, including Dave Abbott, Felicity Ellmore, Shaun Faulkes, Justin O'Riain, Andrew Spinks, Rose Lee, Gonzalo Aguilar, Mark Kirkman, Caroline Rosenthal, David Jacobs, Liz McDaid, David Jarvis and Tim Jackson, to mention but a few. Many past stimulating discussions with Dave Abbott, Andrew Bourke, Mike Bruford, Hynek Burda, Salie Hendricks, Bob Millar, Richard Nichols, Peter Purvis and Justin O'Riain helped to shape some of the ideas contained within this volume.

Marguerite E. Pinenaar and Joelle Wentzel are thanked tremendously for their drawings of mole-rat behavioural acts in Chapters 4 and 5. Tim Jackson is thanked for the stunning photography of the mole-rats. Ingrid Vis kindly typed many of the Tables and Andrew Molteno prepared figures for Chapters 3 and 4.

This book would not have been possible unless our research and that of our cited colleagues had been funded. In this respect we are extremely grateful to the Foundation for Research Development (FRD), the Universities of Cape Town and Pretoria, the Zoological Society of London (Institute of Zoology), the National Geographic Society, the Wellcome Trust, The Royal Society and the Bonhote Bequest of the Linnean Society. For field work, the following people and departments are gratefully thanked: Mike and Eryn Griffin and the Department of Nature Conservation Namibia for their warm hospitality; the National Parks Board, South Africa; Drew Conybeare of the Department of Wildlife Services in Zimbabwe; Mr Nchunga of the Department of Nature Conservation in Botswana; The Office of the President of Kenya and the Wildlife Conservation and Management Department, Kenya; Mr and Mrs H.P. Lühl, Dordabis, Namibia, Mr and Mrs Duckett, Darling, Western Cape and Mr Emerson, Cape Town for allowing us to trap on their respective farms.

Permission to use previously published material was granted by the following organisations and/or journals: Journal of Zoology, London/Zoological Society of London (Chapters 1, 2, 3, 4 and 5) Trends in Ecology and Evolution/Elsevier publishers (Chapters 3 and 8), Oecologia/Springer Verlag (Chapter 3), Physiological Zoology (Chapter 3), Proceedings of the Royal Society Biological Sciences, Series B/ Royal Society (Chapters 6, 8). Comparative Physiology A/ Springer Verlag (Chapter 5).

Finally, we would like to extend our gratitude to our 'small friends with big teeth' for enabling us to share in their lives and allowing us the opportunity to learn so much about their secretive societies.

Chapter 1

Introduction to the Bathyergidae

1.1 BACKGROUND AND HISTORICAL PERSPECTIVES

The phenomenon of eusociality has interested biologists since Darwin first puzzled over, and elegantly explained, how it may evolve by natural selection (Darwin, 1859). Eusociality is a fascinating and special form of social grouping of individuals, and the classical definition was first developed for social insects by Michener (1969) and Wilson (1971). The term 'eusocial' was used to describe groups of animals living in close-knit colonies but where there is a reproductive division of labour. In such social systems, only a small number of individuals are actually involved in direct reproduction. The remainder of the social group is composed of overlapping generations of non-breeding helpers (functionally or irreversibly sterile) that cooperate in the rearing of offspring or maintenance of the colony. With eusocial invertebrates, there is also often a morphological specialisation and differentiation among these different castes (see Wilson, 1971). Until the 1970s, it was thought that eusociality was restricted to invertebrates like the Hymenoptera (bees, wasps and ants) and Isoptera (termites). However, R.D. Alexander, a professor at the University of Michigan working on insect sociality in the mid-1970s, hypothesised on the characteristics that a eusocial mammal would have, should it exist (Sherman *et al.*, 1991). His description fitted that of the naked mole-rat, *Heterocephalus glaber*, and coincident research by J.U.M. Jarvis at the University of Cape Town on the naked mole-rat culminated in a publication by Jarvis (1981), which reported the occurrence of eusociality in a mammal for the first time.

Since the publication of Jarvis's seminal paper, it has become apparent that many vertebrate species exhibit at least some of the key elements of eusociality, namely, a reproductive division of labour and 'helping' behaviour of some kind. Studies of such 'cooperatively breeding' vertebrates, as they are generally known, have flourished, particularly among birds and mammals (see Brown, 1987; Solomon and French, 1997; and Dugatkin, 1997 for recent reviews). The dwarf mongoose, *Helogale parvula* (Rasa, 1977; Rood, 1978), and the meerkat, *Suricata suricatta* (Doolan and Macdonald, 1997) are well-known examples of cooperative breeders and we will return to them in later chapters. Normally, in these species only the dominant female in a group reproduces, and non-breeding helpers assist in anti-predator behaviour and 'babysit' young pups. Among the South American primate family Callitrichidae (marmosets and tamarins), reproduction is again restricted to a single dominant breeding female and non-breeding helpers assist in the carrying of infants, which has a high energetic cost in their arboreal habitat (see Tardif, 1997 and French, 1997 for recent reviews). Clearly, cooperation among animals can encompass a broader range of strategies than these examples of helping within what are essentially kin groups (kin-selected cooperation). Cooperation, in the strict sense of the word, can also include interactions involving acts of reciprocal altruism between non-kin within a species, and also between species. However, it is the former kin-selected altruism with which this book is concerned and readers are referred to Dugatkin (1997) for a recent overview of the subject of cooperation among animals.

Given the wide range of cooperatively breeding strategies among animals, it is important that the semantics, and a generally accepted understanding of exactly what constitutes a eusocial society, are clearly defined. Rigid definitions such as those of Michener and Wilson mentioned at the start of this chapter are not always helpful when taking a broad comparative approach to studies of the evolution of social behaviour in vertebrates, and recently much debate has ensued to resolve these difficulties and establish a more unified approach (see Crespi and Yanega, 1995; Keller and Perrin, 1995; Sherman *et al.*, 1995). This will be discussed more fully in Chapter 8.

Within this broad sphere of research into cooperatively breeding animal societies, the African mole-rats have emerged as a key model system to test hypotheses relating to the evolution and maintenance of sociality. They are arguably a truly unique mammalian group in that they reflect the complete range of social systems from aggressive, solitary-dwelling species, to two species that exhibit the classical features of eusocial societies, the naked mole-rat and the Damaraland mole-rat, *Cryptomys damarensis* (Jarvis 1981; Bennett and Jarvis, 1988a; Jarvis and Bennett, 1993). Apart from their wide array of social systems, African mole-rats have many interesting facets to their biology, mainly as a result of their adaptation to the subterranean niche, as will become evident in some of the following chapters of this book. Naked mole-rats in particular have many unusual traits, not least their 'nakedness', their longevity, with some individuals living over 20 years in captivity (J.U.M. Jarvis, pers. comm.), high levels of inbreeding, and their lack of thermoregulatory ability.

This book aims to explore in detail the range of social systems and the reproductive strategies found in the Bathyergidae, and relate these to their general behaviour and ecology, in an attempt to explore the selection pressures leading to the adoption of a eusocial habit. As we have said, the multiple occurrence of highly social behaviour makes the bathyergid rodents a unique model system to examine the ecological and genetic factors involved in the evolution of cooperative societies, and the physiological processes that maintain them. In comparative analyses of this nature, however, it may not be valid to make comparisons of, for example, sociality and ecological factors, between all species. This is because variables measured in individual species are not always statistically independent points, because traits such as sociality may occur through common descent, rather than evolving independently (Harvey and Pagel, 1991). Detailed knowledge of phylogenetic (evolutionary) relationships between taxa is therefore essential, and molecular genetics techniques such as DNA sequence analysis are a powerful tool in phylogenetic analyses, particularly among cryptic or convergent species. This is especially important for the Bathyergidae as taxonomy of the family at a number of levels has been confused. Our current knowledge of the systematics and distribution of the Bathyergidae will be the subject of the rest of this chapter.

1.2 THE FAMILY AND ITS SYSTEMATICS

Traditionally, the order Rodentia is divided into three sub-orders, the Myomorpha (mouse-like), the Sciuromorpha (squirrel-like) and the Hystricomorpha or Hystricognathi (cavy-like). The placement of taxa within these three groups is determined by morphometric parameters including the arrangement of jaw muscles and shape of the skull (Wood, 1985), in particular the position and attachment of the masseter muscles to the skull and lower jaw. While, in the past, these criteria did not allow the bathyergids to be unambiguously assigned to a sub-order, they appeared to be closest to the Hystricomorpha (for review see Honeycutt *et al.*, 1991a). Certain characteristics of the reproductive system of mole-rats, such as ovarian cycle length, relatively long gestational periods and structural and functional modifications of the ovary in the form of structures called accessory corpora lutea (Faulkes *et al.*, 1990a; Chapter 5), also support their affinity with hystricomorphs. Morphometric synapomorphies (shared derived characters indicating common ancestry) further suggest that the Bathyergidae are a monophyletic group, i.e. all taxa can be traced back to a single common ancestor (Wood, 1985).

More recently a comprehensive molecular phylogeny, based on sequence differences of the mitochondrial 12S rRNA gene, that included 22 representatives of 14 of the 16 families, has helped to elucidate further the evolutionary relationships of the hystricomorph rodents (Nedbal *et al.*, 1994). In recent years these methods have contributed greatly to the understanding of phylogeny in many groups (Avise, 1994) and are based on the assumption that sequence differences between lineages accumulate over time as a result of mutations. Therefore, the greater the amount of sequence difference, the more divergent the sequences (or individuals) are, and the longer ago it was since a shared common ancestry. This analysis, and other morphological data, strongly supports a monophyletic origin for the New World 'caviomorpha' families which are mainly found in South America, and has good support for monophyly of the Old World 'phiomorpha' families, and monophyly for the Bathyergidae (Figure 1.1). It is in the phiomorph (Old World) grouping that we find the Bathyergidae with their closest relatives, the rock rats (family Petromuridae), the cane rats (family Thryonomyidae) and the Old World porcupines (family Hystricidae). Using the 12S rRNA sequence differences as a 'molecular clock' to estimate divergence times of the hystricomorph families, Nedbal *et al.* (1994) calculate that the New World caviomorpha and Old World phiomorpha separated from their common ancestor during the Eocene, approximately 33–39 million years ago. However, difficulties in calibrating the molecular clock accurately led Nedbal *et al.* to extend the upper limit of the range to 49 million years ago. Thus, the

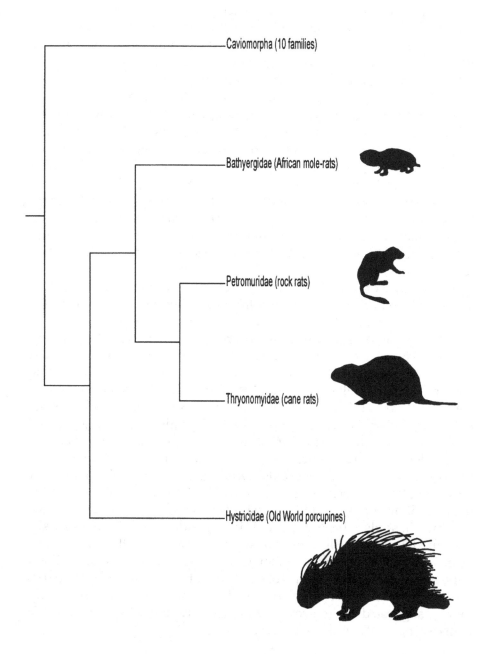

Figure 1.1. Phylogenetic relationships of the four Old World phiomorph families and the New World caviomorpha in the Hystricomorph sub-order of rodents, based on parsimony analysis of sequence differences in the mitochondrial 12S rRNA gene (adapted from Nedbal *et al.*, 1994).

hystricognath rodents have an ancient origin within the class Mammalia, and recent molecular analysis of 52 genes suggests that the lineage may have diverged from other mammals as long ago as 109 million years, during the mid-Cretaceous, and before the extinction of dinosaurs (Kumar and Hedges, 1998).

Within the Bathyergidae, both Roberts (1951) and De Graaff (1981) suggest a grouping of the five genera in the family into two sub-families based on dental characteristics. The sub-family Bathyerginae contains a single genus, *Bathyergus*, whose species are characterised by grooved upper incisors and ungrooved lower incisors, large body size (up to 1800 g) and enlarged forefeet and claws. The other four genera of the family, *Heterocephalus*, *Heliophobius*, *Georychus* and *Cryptomys*, are included in the sub-family Georychinae, characterised by ungrooved incisors, more delicate un-enlarged forefeet and claws and body sizes generally less than 400 g (De Graaff, 1981).

Knowledge of the palaeontology of African mole-rats is rather limited, but has been reviewed by Jarvis and Bennett (1990). From the limited data available, the Bathyergidae are thought to have an ancient African origin, and fossils from three genera have been found in early Miocene fossil beds (around 25 million years ago) in East Africa and Namibia (Lavocat, 1973, 1978). This is also in keeping with molecular phylogenetic studies (Allard and Honeycutt, 1992; Nedbal *et al.*, 1994). The largest of these three fossil genera, *Bathyergoides*, while distinct from the Bathyergidae, is thought to be related to the latter. Another of the fossil genera, *Proheliophobius*, morphologically resembles some of the extant Bathyergidae (*Heterocephalus* and *Heliophobius*), while part of a skull of another fossil species, *Paracryptomys*, was found in the early Miocene beds of the Namib desert in Namibia. Fossil evidence of *Heterocephalus* in the Miocene has also been reported in Uganda, with further fossils described from the Pliocene/Pleistocene (from 7 million to 10,000 years ago) in East Africa (Van Couvering, 1980). The coincident appearance of *Heterocephalus* fossils with extinct bathyergid ancestors also supports the early divergence of *Heterocephalus* within the family that is suggested by molecular data (Allard and Honeycutt, 1992; Faulkes *et al.*, 1997a; this chapter).

century (*Bathyergus suillus*), and since then a plethora of type specimens have been recorded, particularly around the late nineteenth and early twentieth centuries. Many of the original specimens were caught by what were probably amateur naturalists, and the accompanying notes are often embellished with fascinating and trivial details. Good examples of this are the notes by De Winton (1896), who examined specimens of *Georychus darlingi* (a synonym for *Cryptomys darlingi*) and the field notes of Mr Darling, who had made a collection of small mammals while engaged in mining work in Mashunaland (now Zimbabwe): 'Every specimen has been most carefully prepared, with date of capture, sex and measurements, taken in the flesh, recorded, and in almost every case the skull accompanies the skin; in the few cases when this is missing, it is fully accounted for by having been eaten by a hen or some evil beast; in one case the skull alone is sent, the skin preserved with arsenical soap and stuffed with cotton wool, having been eaten and vomited by a cat, was thought not to be worth postage, which, by the way, is 2s. 9d. per lb.' These carefully recorded specimens now residing in collections are a valuable potential source of genetic material, with the advent of molecular techniques for extracting and analysing DNA from ancient and modern museum specimens.

The Bathyergidae are endemic to sub-Saharan Africa and occur in a wide range of physically and climatically divergent habitats (mesic to xeric). They can excavate their burrows in a wide range of soil types ranging from coarse arenosols (sands) to fine clays. They also can be found over a wide range of altitudes, rainfall patterns and vegetation types. One common denominator in the distribution of mole-rats above all other factors is the presence of geophytes (roots, corms and tubers) that form the staple diet of all species (see Chapter 3). The two species within the genus *Bathyergus* also consume the aerial vegetation of plants. Mole-rats do not drink, and obtain all their water and nutritional requirements from their food.

Some mole-rats are strictly allopatric or parapatric in their distributions, whereas others occur in sympatry, at least in part of their distributional range. Of the solitary species, three genera are found in mesic habitats, these being *Bathyergus*, *Georychus* and *Heliophobius*. The solitary mole-rats are generally larger and restricted to regions of higher precipitation (>400 mm per annum). In contrast, the two social genera (*Cryptomys* and *Heterocephalus*) are characteristically smaller than their solitary counterparts and are found in both mesic and xeric regions. It should be noted, however, that there are some exceptions to the general rule. The social genera

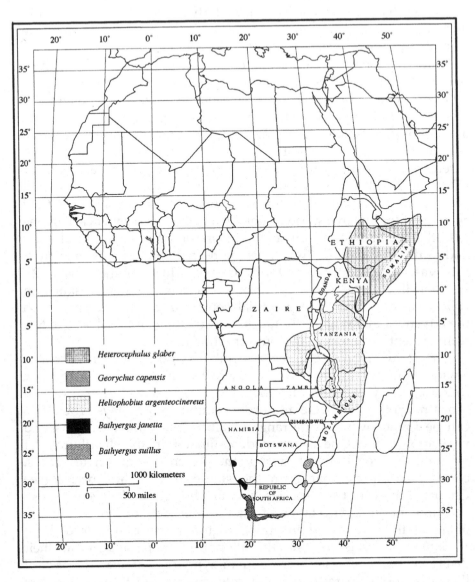

Figure 1.2. Distribution map for the naked mole-rat, *Heterocephalus glaber*, and four species of solitary bathyergid mole-rats.

are particularly successful and generally have wider distributional ranges than the solitary species (Jarvis and Bennett, 1990, 1991). The ranges of some of these social species extend into areas of very low rainfall, which is sporadic and unpredictable (sometimes <200 mm per annum). The relationships between rainfall patterns, food distribution and sociality are discussed fully in Chapter 8.

The sub-family Bathyerginae comprises a single genus, *Bathyergus*, containing two species, the Cape dune mole-rat, *Bathyergus suillus*, and the Namaqua dune mole-rat, *Bathyergus janetta*, endemic to south and southwestern Africa (Figure 1.2). The Cape dune mole-rat (Schreber, 1782) is a solitary species restricted to the coastal area of the southwestern Cape Province from Lamberts Bay and Klawer in the northwest, to Knysna on the south coast, penetrating inland for approximately 80 km in the northwest. It occurs at low altitudes inhabiting areas below 300 m above sea level (De Graaff, 1981). The Cape dune mole-rat is predominantly associated with sand dunes but can also be found in coarse sandy loams and arenosols on sand flats or in alluvial sand along river systems. These areas are characterised by high annual rainfall (over 500 mm). It is the largest species within the family, and has a sandy/buff coloured pelage and a white head-patch.

The Namaqua dune mole-rat (Thomas and Schwann, 1904) is also a solitary species. It is confined to the northwestern parts of the Cape Province in coastal sand dunes, as far north as Oranjemund, and northwards into the coastal diamond areas of Boegoeberg and Bosmanberg of Namibia, approximately 90 km north of the Orange River (J.U.M. Jarvis and M. Griffin, pers. comm.). In the southern part of their distribution, they extend as far inland as Steinkopf and Ezelfontein in the Kamiesberg at an altitude of 1350 m. In their coastal range, they are usually found in sand dunes and inland in a substrate of sandy alluvium. The Namaqua dune mole-rat is of particular interest in studies of the evolution of sociality because it is a solitary species living in an arid habitat (normally the solitary species are restricted to mesic areas, and sociality is thought to be linked to aridity). In fact, at one field site near Steinkopf, South Africa, rainfall figures are amongst the lowest in Africa at less than 100 mm per annum. However, in this habitat, rainfall data alone can be misleading, as these areas gain additional moisture in the form of underground seepage areas (J.U.M. Jarvis, pers. comm.).

The Namaqua dune mole-rat is smaller than the Cape dune mole-rat, usually no more than 230 mm in total head and body length. It has grey to buff pelage dorsally, a grey-brown band along the dorsal mid-line, with a nearly black head, and pronounced white areas around the eyes, muzzle and ears (Honeycutt *et al.*, 1991a).

The ranges of the Cape dune mole-rat and the Namaqua dune mole-rat show no overlap but exhibit sympatry with the common mole-rat, *Cryptomys hottentotus hottentotus*. The Cape dune mole-rat

Figure 1.3. The Cape dune mole-rat, *Bathyergus suillus* (photograph by Jenny Jarvis).

may also exhibit limited sympatry with the Cape mole-rat, *Georychus capensis*. The dune mole-rats are the largest of the bathyergids, with a body size up to 2000 g in large males of the Cape dune mole-rat (Figures 1.3 and 1.4a). They exhibit sexual dimorphism based on body mass, the males being significantly larger than the females (Davies and Jarvis, 1986). For example, mean body masses for the Cape dune mole-rat are 933 g (males) versus 635 g (females), and for the Namaqua dune mole-rat, 451 g (males) versus 332 g (females) (Jarvis and Bennett, 1991). Dune mole-rats differ from the other species in that they utilise their clawed forefeet, as well as their incisors, for excavating the soil.

The sub-family Georychinae comprises four genera, *Georychus*, *Heliophobius*, *Heterocephalus* and *Cryptomys*, and has a wide geographical distribution with species occurring in western, central, eastern and southern Africa. The Georychinae occur in more compact soils when compared to their Bathyerginae counterparts, and species are generally smaller and rarely have a body mass exceeding 600 g (Wallace and Bennett, 1998). Species in the genera *Cryptomys* and *Heterocephalus* are social. The species occurring in these two genera are found in colonies where there is a reproductive division of labour and a sole reproductive female in each colony is involved in breeding. In

(a)

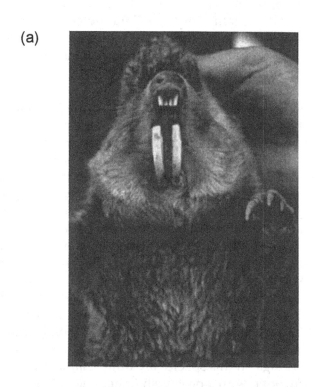

(b)

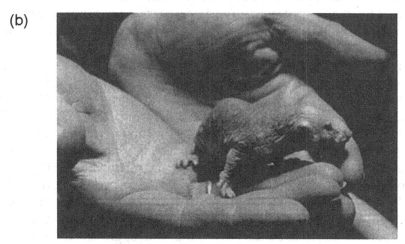

Figure 1.4. (a) The largest member of the Bathyergidae, the Cape dune mole-rat, *Bathyergus suillus*; males may reach up to 2 kg, and (b) the smallest, the naked mole-rat, *Heterocephalus glaber* averages just over 30 g. The well-developed claws characteristic of the Bathyerginae sub-family are clearly visible (photographs by Chris Faulkes and Terry Dennett).

the genus *Cryptomys* colonies of up to 41 have been reported (Jarvis and Bennett, 1993). Colonies of the naked mole-rat may number up to 295 individuals, although 90 or more individuals is more common (Brett, 1986). The other two genera, *Georychus* and *Heliophobius*, are strictly solitary and plural occupancy of a burrow occurs only briefly during the breeding period.

The Cape mole-rat, *Georychus capensis*, was the first bathyergid to be recorded in the literature (Pallas, 1778). It occurs in areas that have, on average, over 500 mm rainfall per annum, i.e. in the Cape Province from the Cape Peninsula northwards into Niewoudtville, inland to Worcester and Tulbagh, eastwards into the Transkei and in the montane regions of the southwestern Cape Province (Figures 1.2 and 1.5). There are no documented sightings or recordings between these areas and isolated populations in southwestern Natal near the border of Lesotho and Belfast and Ermelo in the Transvaal (Figure 1.2). It is possible that these disjunct populations could represent two species as allozyme and mitochondrial DNA analysis suggest they are divergent (Honeycutt *et al.*, 1987; Nevo *et al.*, 1987). These solitary mole-rats are found in sandy loams and alluvium and also clay soils. The Cape mole-rat is not commonly found in sand dunes or the coarse beach sands frequented by the Cape dune mole-rat. The pelage and markings on this species are quite characteristic. The body is a buff-orange, often with a brownish tinge, and a black to charcoal coloured head and contrasting white patches on the muzzle, ears and eyes. Body size is around 180 g for both sexes (Jarvis and Bennett, 1991).

The silvery mole-rat, *Heliophobius argenteocinereus*, has a wide range and occurs in the sandy soils of savannas and woodlands of southern Kenya, throughout Tanzania and parts of southeastern Zaire, to central Mozambique (Figures 1.2 and 1.6; Peters, 1852). These areas are characterised by a high annual rainfall which on average exceeds 900 mm. The genus may be monotypic, although in the past a second species, *Heliophobius spalax* (Thomas, 1910), found at Taveta and the Taita Hills of Kenya, has also been recognised (De Graaff, 1971). Honeycutt *et al.* (1991a) re-examined specimens of both types and concluded that differences were due to age variation. The silvery mole-rat is solitary dwelling and aggressive. There is no apparent sexual dimorphism, with adult animals weighing an average of 160 g (Jarvis, 1973a). The pelage is pale sandy, reddish or greyish dorsally, with slightly paler underparts. Apart from this, comparatively little is known about its ecology or behaviour in the wild.

Figure 1.5. The Cape mole-rat, *Georychus capensis*, showing characteristic white spectacle-like eye markings (photograph © Tim Jackson).

Figure 1.6. The silvery mole-rat, *Heliophobius argenteocinereus* (photograph by Jenny Jarvis).

The naked mole-rat, or sand puppy, *Heterocephalus glaber*, is found in the arid regions of East Africa, from the Rift Valley of Ethiopia eastwards into the north of Somalia, and from Lake Turkana in Kenya eastwards to the coast of Somalia, and south as far as Tsavo National Park in Kenya (Figures 1.2, 1.4b and 1.7) (Rüppell, 1842). It is generally found in areas with fine sandy soils which become hard in dry seasons, only softening after rain (Brett, 1991a). These areas are characterised by low (less than 400 mm per year) and unpredictable rainfall, with on average only four months per year having more than 25 mm of rain. It has been shown by Jarvis *et al.* (1994) that 25 mm is approximately the quantity required to soften the soil at the depth of foraging tunnels and facilitate burrowing.

Figure 1.7. The naked mole-rat, *Heterocephalus glaber* (photograph © Neil Bromhall).

Naked mole-rats have a unique appearance (Figures 1.4b and 1.7) and are the smallest of the bathyergids with a mean body mass of 34 g (Jarvis and Bennett, 1991). When one was first caught and named by Rüppell in 1842, he thought that he had a diseased or decrepit specimen, and the name *Heterocephalus glaber* loosely translated means animal with a smooth skin and oddly shaped head. It wasn't for a number of years, until 1885, when another specimen was examined by Oldfield Thomas, that it was realised that Rüppell's animal was 'normal'. In a letter to Oldfield Thomas, E. Lort

Phillips described the animal as a '... sort of mole ... tail like that of a hippo ... teeth like those of a walrus.' While naked mole-rats lack a pelage, they do have a covering of sensory vibrissae which are immediately apparent if the animal is held up to the light. The naked skin is puzzling, considering all the other members of the family have a normal pelage, and possible reasons for their virtual hairlessness have been discussed by Alexander (1991).

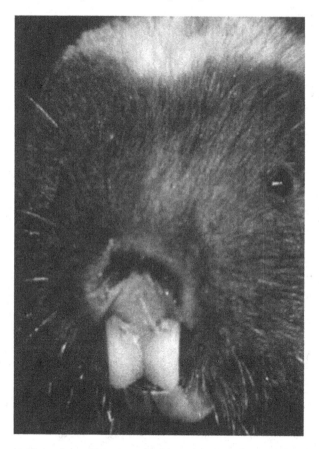

Figure 1.8. The Damaraland mole-rat, *Cryptomys damarensis* (photograph © Tim Jackson).

The Damaraland mole-rat, *Cryptomys damarensis*, has a wide distribution (Ogilby, 1838), occurring in the extreme western parts of Zambia, western Zimbabwe (Bulawayo district), Botswana (Maun, Chobe and Moremi), the Caprivi strip, central and northern Namibia (Rheoboth to Etosha) and the northern and northwestern Cape Province (Hotazel to Twee Rivieren and Nossob). It occurs

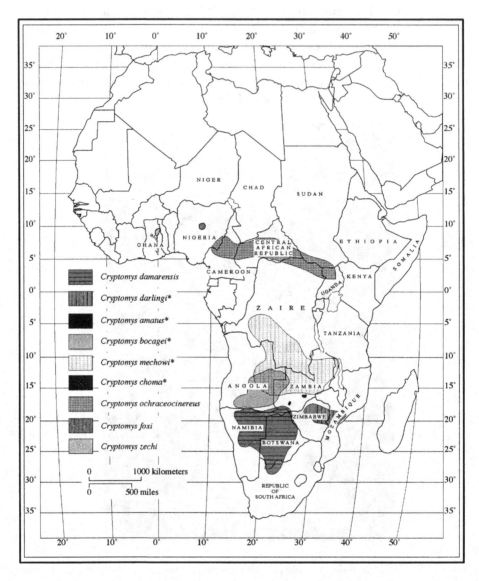

Figure 1.9. Distribution map for *C. damarensis* and subspecies. * denotes approximate ranges only are given for these taxa, and sympatry may occur.

predominantly in the red Kalahari arenosols and in loose unconsolidated alluvial sands. The dune slopes and valleys of the Kalahari sand complex are the type locality for this species (Figures 1.8 and 1.9). Like the naked mole-rat, the Damaraland mole-rat generally inhabits areas of low and unpredictable rain with a mean

annual rainfall averaging less than 400 mm and, on average, only four
months a year exceeding 25 mm rainfall. After the naked mole-rat,
the Damaraland mole-rat has the highest maximum group size at 41,
with an average of 16 per colony and, like the naked mole-rat,
reproduction is restricted to a single breeding pair. Mean body mass
is 131 g (Jarvis and Bennett, 1993). There is some variation in the
pelage colour with two colour morphs, fawn and dark brown/black,
which can occur within one colony. In the southern part of their
distribution, the darker form predominates, and both colour morphs
have a light head patch.

Figure 1.10. The giant Zambian mole-rat, *Cryptomys mechowi* (photograph by Chris
Faulkes).

The Mashona mole-rat, *Cryptomys darlingi*, was first described by
Thomas (1895). It is found in scrub habitats and Miombo woodland
in northern and eastern areas of Zimbabwe and is believed to extend
into western Mozambique. Burrows interdigitate between the roots of
trees and shrubs. To date, little is known about its full distributional
range, except that it appears to be restricted to the eastern sections of
Zimbabwe and west and central Mozambique (Figure 1.9). The
Mashona mole-rat occurs in sandstone and granitic-derived soils in
areas of relatively high, predictable rainfall (mean annual rainfall over
700 mm). Colonies contain five to nine individuals with one breeding
pair and average body size for both sexes is around 64 g (Bennett *et*

al., 1994a). Pelage colour ranges from blackish through seal brown to slate and silvery-fawn, and a white head-patch is apparent (De Graaff, 1971).

The giant Zambian mole-rat, *Cryptomys mechowi* (Peters, 1881) lives in diverse soil types in the mesic Miombo tropical woodland and savanna of northern Zambia, southern Zaire and central Angola (Figures 1.9 and 1.10; De Graaff, 1971). Rainfall in these areas is relatively high at over 1100 mm per annum. It is pale brown in colour, and the head-spot characteristic of some species in this genus is absent. Group sizes for this species in the wild remain unclear although Burda and Kawalika (1993) report catching eight individuals in one burrow, but also state that, according to hunters, colonies may contain over 60. This seems unlikely, given that litters only number one to four pups (usually two or three), and gestation is around 100 days (Bennett and Aguilar, 1995). Assuming that only one pair breeds per colony, it would take at least 10 years for groups of this size to form, assuming no deaths, immigration or emigration. Two to six animals per colony ($n = 4$) were caught in northern Zambia by Faulkes *et al.* (1997a) while De Graaff (1971) reports that the giant Zambian mole-rat is usually solitary. Thus, the consensus tends to lean towards small group sizes, although more work is required on this species to corroborate the anecdotal reports of large colonies. Body size is relatively large and sexual dimorphism is apparent, with males up to 600 g and females up to 350 g (Burda and Kawalika, 1993).

In addition to the giant Zambian mole-rat, there are a number of other divergent populations of Zambian mole-rats of unclear taxonomic status. *C. amatus* (Wroughton, 1907), has previously been described as a subspecies of *C. hottentotus*, but is clearly more closely related to the Damaraland mole-rat (see Section 1.4). Little is known about *C. amatus* and its distribution is uncertain. The type locality is described from specimens collected from the Alala Plateau in northeastern Zambia, but it also apparently occurs in and around Lusaka, where colonies were found to have a maximum group size of 10 with one breeding pair. It is a relatively small mole-rat at around 67 g body mass, and the pelage has a characteristic russet tinge with a white head-patch (Faulkes *et al.*, 1997a; C.G. Faulkes, N.C. Bennett and J.U.M. Jarvis, unpubl.). Another two Zambian taxa having different karyotypes, one of $2n = 58$ found at Itezhi-Tezhi (Filippucci *et al.*, 1994, 1997) and another of $2n = 50$ at Choma (Faulkes *et al.*, 1997a), require further investigation.

The Angolan mole-rat, *Cryptomys bocagei* (De Winton, 1897) is found in western Angola, ranging southwards into northern Namibia. Little is known of its habits in the wild. Rainfall across its range tends to be high, at over 800 mm per annum. Colony sizes are probably small, with four being caught in one colony (J.U.M. Jarvis and G.H. Aguilar, pers. comm.; Figure 1.9). Pelage is cinnamon to drab, with the variable presence of a head-spot and body size is small – up to 150 mm head and body length, cf. the giant Zambian mole-rat, which can measure up to 260 mm (Ellerman, 1940). Very little is known of the distribution and ecology of the disjunct species of *Cryptomys* occurring north of the equator. Honeycutt *et al.* (1991a) recognise three species although, as with other cryptomids, a larger number of different forms have been described in the past. The Nigerian mole-rat (*Cryptomys foxi*) is found in northern Nigeria (Figure 1.9) and possibly into Cameroon (Thomas, 1911) where a number of specimens referred to as *C. foxi* were caught by Williams *et al.* (1983). Nothing is known of their social structure, although, of the 12 original specimens caught by the Rev. George Fox, after whom the species is named, nine were morphologically similar (adults), while one male and one female had much longer bodies and more massive skulls (cited in Rosevear, 1969). If these were from one colony (and this is not stated), then these samples could represent a larger breeding pair and their non-breeding colony members. Body size is small, with head and body length up to 160 mm (mean 145 mm), and the pelage black with a variable white head-patch and occasionally white ventral markings (Rosevear, 1969; Williams *et al.*, 1983).

The Togo mole-rat, *Cryptomys zechi* is reported in northeastern Ghana and northwestern Togo (Matschie, 1900) although its exact distribution is at present unclear (Figure 1.9). Pelage is 'pallid' but variable in colour from buff to light cinnamon drab. There is a variable presence of head-spot and mean head and body length is 167 mm (Rosevear, 1969).

The ochre mole-rat, *Cryptomys ochraceocinereus* (Heuglin, 1864), has an uncertain distribution. The original specimens were described in Sudan but may also include as synonyms *C. lechei* from Uganda and northeastern Zaire and *C. kummi* from Central African Republic (Figure 1.9). Head and body length is up to 200 mm (Rosevear, 1969) and pelage is brown to sepia, with a white head-patch.

The common mole-rat, *Cryptomys hottentotus hottentotus*, and its subspecies have a wide distribution (Lesson, 1826; Figures 1.11 and 1.12) and are found sympatrically with all mole-rat species in South Africa apart from the Damaraland mole-rat. They are found as far south as the northern suburbs of the Cape Peninsula, extending

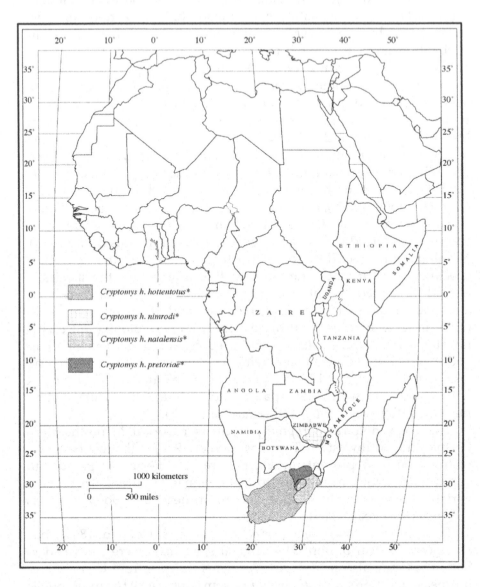

Figure 1.11. Distribution map for *C. h. hottentotus* and related species. * denotes approximate ranges only are given for these taxa, and sympatry may occur.

northwards to Steinkopf and inland through to Prieska and Calvinia, and southwards around the coast and inland through the eastern Cape, Natal and the western and northern Transvaal. The common mole-rat is a particularly interesting species because it has a wide distribution and occurs within a range of different rainfall patterns.

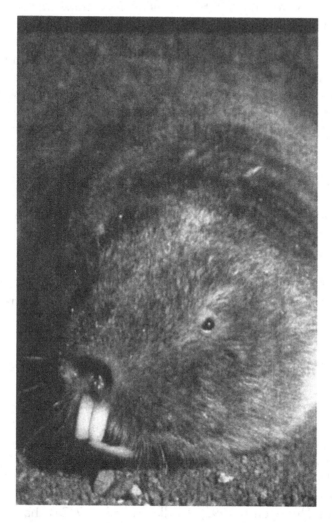

Figure 1.12. The common mole-rat, *Cryptomys hottentotus hottentotus.* Note the absence of a head-patch (photograph © Tim Jackson).

Common mole-rats also occur in a wide diversity of substrates ranging from sandy loams through to heavier, more compact soils such as exfoliated schists and stony soils, although they are not fond of heavy clay soils or very brecciated soils. In general, they tend to prefer granitic soils, in addition to sandy alluvium along river systems. The common mole-rat is small with a mean body mass of 77 g for males and 57 g for females (Jarvis and Bennett, 1991). The pelage is drab or fawn, occasionally blackish with no head-patch

(Ellerman, 1940).

The Matabeleland mole-rat, *Cryptomys hottentotus nimrodi*, occurs in groups of eight to 15 individuals, with reproduction restricted to a single male and female, and is found in the red karroo arenosols and granitic loams of southeastern Zimbabwe (Figure 1.11; Bennett *et al.*, 1996a). Relatively little is known about this species.

The Natal mole-rat, *C. h. natalensis* and *C. h. pretoriae* (sometimes referred to as the highveld mole-rat) have average group sizes of eight and 12 respectively, with a single breeding pair per colony. Mean body size is 106 g for males and 88 g for females (Jarvis and Bennett, 1991). Although the Natal mole-rat and the common mole-rat both have a 2n of 54, it is unclear whether they constitute separate species (see Section 1.4).

The above list is not exhaustive, and there are other taxa of *Cryptomys* that require further investigation. These include, for example, *C. whytei* (Thomas, 1897), found in Karonga at the northwest corner of Lake Nyasa in northern Malawi, which appears to be somewhat separate in its range from the other *Cryptomys* species described here, and populations in Zambia and central Africa already mentioned.

1.4 PHYLOGENETIC RELATIONSHIPS WITHIN THE FAMILY

Beyond classification at the sub-familial level, relationships within the family have also been the subject of much debate. In a recent review of the literature, Honeycutt *et al.* (1991a) suggest that five genera and 12 species are recognised, based mainly on morphological characteristics. Following on from this, molecular genetic studies (Allard and Honeycutt, 1992; Faulkes *et al.*, 1997a) have enabled genetic differences between species and the numerous proposed sub-species to be quantified and their relationships clarified. It is apparent from this that a revision of the current classification is required, mainly within the genus *Cryptomys*. Figure 1.13 shows the most current and complete molecular phylogeny of the family available to date, based on analysis of differences in the sequence of bases of the mitochondrial 12S rRNA and cytochrome-*b* genes (Allard and Honeycutt, 1992; Faulkes *et al.*, 1997a). When the mole-rat sequences were analysed using a number of different methods to reconstruct the evolutionary trees, all these methods produced trees having the same overall topology with only slight differences in the grouping of the solitary species relative to one another.

The phylogram displayed in Figure 1.13 shows the three genera containing four solitary living species grouped independently to the two social genera *Heterocephalus* and *Cryptomys*. The monotypic genus containing the naked mole-rat is ancestral in the family and divergent from the Damaraland mole-rat, suggesting the parallel evolution of eusociality within the family (Allard and Honeycutt, 1992; Faulkes *et al.*, 1997a).

In contrast to the other genera within the family, the genus *Cryptomys* is species-rich and, in the past, approximately 49 forms have been named (Ellerman, 1940). Figure 1.13 contains 10 taxa that exhibit a variety of levels of sociality, and they divide into two distinct sub-clades within the tree, one containing the Damaraland mole-rat, *C. damarensis* and one containing the common mole-rat, *C. h. hottentotus*. Taxa in these two sub-clades of *Cryptomys* have minimum genetic distances between them of the same magnitude as inter-generic distances within the family. The genetic distances obtained from calculating the percentage sequence differences of mitochondrial DNA between individuals enables quantitative comparisons to be made between taxa.

The data obtained from an analysis of cytochrome-*b* percentage sequence differences in mole-rats are summarised in Figure 1.14 (Faulkes *et al.*, 1997a). For example, for cytochrome-*b*, comparison of *C. h. hottentotus* with *C. bocagei* gave differences of 20.6%, compared with *Bathergus janetta* versus *Georychus capensis*, where differences numbered 21.9% (Figure 1.14). Whether or not these two sub-clades should be considered as separate genera is a matter for debate, but clearly they are divergent, and this may only be resolved by further analysis of morphological data in the light of these molecular results. Indeed, the limited morphological data available are supportive of these patterns of divergence. When the origin and insertion of the masseter muscles of the jaw are examined relative to the infraorbital foramen, zygomatic arch and lower jaw, *C. mechowi*, *C. damarensis* and *C. bocagei* all have thick-walled infraorbital foramina, while *C. h. hottentotus* has an infraorbital foramen that is elliptically shaped and has a thin external wall (Honeycutt *et al.*, 1991a; G.H. Aguilar, pers. comm.).

The smallest genetic distances we have so far obtained between currently recognised species are 8.37% for *C. mechowi* and *C. bocagei*, and 5.4% for *B. suillus* and *B. janetta* (cytochrome-*b*). Mean percentage within species genetic differences (summarised in Figure

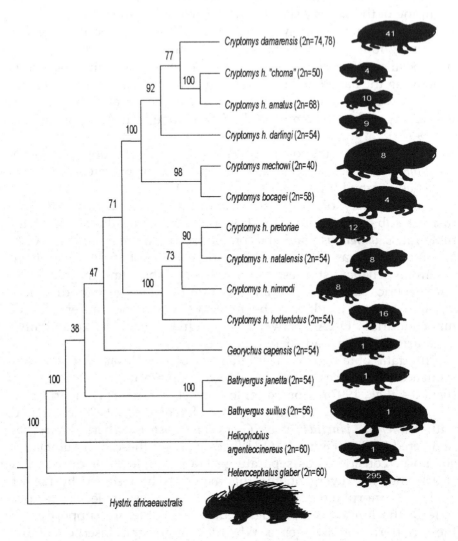

Figure 1.13. Phylogenetic relationships of 15 African mole-rat haplotypes and one outgroup species (*H. africaeaustralis*) based on parsimony analysis of sequence differences in the mitochondrial 12S rRNA and cytochrome-*b* genes. This is a consensus of two trees produced by the analysis which differed only in the branching order of *Bathyergus* and *Heliophobius*. Maximum social group sizes are shown on the schematic of each species, which are drawn approximately relative to body size. Diploid numbers (2n), where known, are shown in parentheses. Numbers above branches are bootstrap values obtained after resampling the data set 100 times, and indicates the % number of times that the respective node was supported (internal nodes are the branching points that represent common ancestors (adapted from Faulkes *et al.*, 1997a).

1.14; also see Chapter 6) have been quantified for cytochrome-*b* within and between geographic locations for three species as follows: *H. glaber* (for 10 individuals), within 1.50, between 4.44 (range 2.00–5.80); *C. damarensis* (10 individuals), within 0.86, between 0.78 (range 0.09–1.47); *C. h. hottentotus* (seven individuals), between 3.89 (range 0.50–7.05). Using these values as an approximate index, and with reference to the phylogenetic relationships displayed in Figure 1.13, it is clear that *C. amatus* from Lusaka and *C. darlingi* from Harare, traditionally classified as subspecies of *C. hottentotus*, are divergent and should be classed as separate species. These molecular data support a previous suggestion, based on karyotypic differences in the fundamental number, that *C. darlingi* should be regarded as a distinct species (Aguilar, 1993). A species of cryptomid caught near Choma in Zambia, which grouped with *C. amatus* but was karyotypically quite different (Figure 1.13) and had 3.42% difference at cytochrome-*b* (within our observed intra-specific range), should possibly be considered as a sub-species of *C. amatus*. Other populations of Zambian mole-rats are also known to be divergent, with another form having 2n = 58 found at Itezhi-Tezhi (Filippucci *et al.*, 1994; 1997). Within the *C. hottentotus* sub-clade (Figure 1.13), *C. h. nimrodi* from Zimbabwe, *C. h. natalensis* from Natal and *C. hottentotus* captured near Pretoria (referred to as *C. h. pretoriae*) were divergent with minimum cytochome-*b* distances of 8.45% between them, and a minimum of 11.45% compared with *C. h. hottentotus*, and should perhaps also be considered as separate species (Faulkes *et al.*, 1997a).

Three other geographically distinct species of *Cryptomys*, *C. foxi* (Thomas, 1911), *C. zechi* (Matschie, 1900) and *C. ochraceocinereus* (Heuglin, 1864) were not included in the sequence analysis as sufficient data were not available. However, preliminary sequence analysis of DNA extracted from museum specimens places these three species in the same clade as *C. damarensis* (C.G. Faulkes, unpubl.). Morphological affinities in support of this have also been noted previously: *C. zechi*, *C. foxi* and *C. ochraceocinereus* all have the small and thick-walled infraorbital foramina seen in *C. mechowi*, *C. damarensis* and *C. bocagei* (Honeycutt *et al.*, 1991a).

While one might expect a degree of synonymy within the genus *Cryptomys*, given that 49 forms have been named in the past (Ellerman, 1940), recent molecular genetic studies suggest that there are more species than the seven currently recognised by Honeycutt *et al.*'s (1991a) review based on morphological characteristics. Although there are no established criteria for the assignment of subspecies,

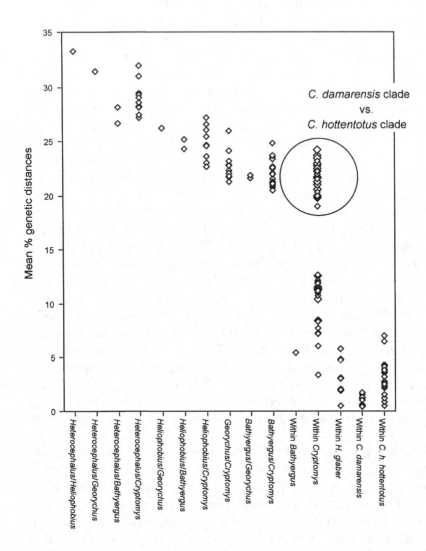

Figure 1.14. Mean percentage genetic differences (calculated from Kimura's 2-parameter model) for pairwise comparisons of mitochondrial cytochrome-*b* DNA sequences within and between genera and species for the family Bathyergidae. The circled values refer to pairwise comparisons between individual species of the two *Cryptomys* sub-clades shown in Figure 1.13.

species or genera from the molecular data described earlier, we can make valid comparisons of sequence divergence within the family. What constitutes a species is dependent on which definition is applied and has been the subject of much debate. The biological species concept (Mayr, 1942) states that species are reproductively isolated populations, while the wider-ranging phylogenetic species concept (Cracraft, 1983) defines a species as a group containing one diagnostic character present in all individuals, but absent in close relatives of the group. No information is available at present as to whether hybridisation of various closely related taxa is possible (in captivity), e.g. *C. bocagei* and *C. mechowi*, which would help to further clarify the taxonomy in terms of a biological species concept approach.

The *Heterocephalus* genus is considered to be monotypic, although a number of different taxa were named around the turn of the century (see Honeycutt *et al.*, 1991a for review). Mitochondrial DNA sequence analysis has revealed mean genetic differences between Ethiopian and southern Kenyan (Mtito Andei) haplotypes of 5.8% for the cytochrome-*b* gene (see Chapter 7). Assuming an approximate divergence rate of 2% per million years for the mole-rat cytochrome-*b* gene, then this would put the time since the Ethiopian and Mtito Andei individuals shared a common ancestor at approximately 2.9 million years ago. The fossil record of naked mole-rats in east Africa is known to extend back approximately three million years (Van Couvering, 1980), suggesting an ancient divergence between Ethiopian and Kenyan populations. The value of 5.8% divergence between Ethiopian and southern Kenyan haplotypes is slightly greater than the intra-specific range for cytochrome-*b* genetic distances known for other African mole-rats (see Chapter 7) and approaching that of some inter-specific distances (for example, *Bathyergus suillus* versus *Bathyergus janetta*, 5.9% and *Cryptomys damarensis* versus *Cryptomys amatus*, 6.6%). Hamilton (1928) noted morphological differences between naked mole-rats he examined from Kenya and those caught in Somalia by Parona and Cattaneo (1893). These differences include the presence of 10 pairs of ribs, with an eleventh pair vestigial in the Kenyan animals, while the Somali mole-rats of Parona and Cattaneo had 11 pairs of ribs, and a twelfth pair rudimentary. Whether these widely dispersed naked mole-rat populations, which are highly divergent, are distinct species is unclear and again depends on species definition. Animals from northern Kenyan colonies at Lerata will readily breed with southern Kenyan Mtito Andei individuals in captivity, despite genetic distances of 3% between them at the cytochrome-*b* gene. It is at present not known whether this would also hold true for pairs of Kenyan and Ethiopian or Somalian individuals, and unfortunately all captive naked mole-rats are currently derived from Kenyan stock (Faulkes *et al.*, 1997b).

1.5 PHYLOGEOGRAPHIC RELATIONSHIPS

If the phylogeny in Figure 1.13 is mapped onto the distribution patterns of the extant species of mole-rats, some phylogeographic patterns emerge. These have been discussed previously by Allard and Honeycutt (1992) and their conclusions were dependent on the exact placement of the silvery mole-rat within the phylogeny. The position of *Heliophobius* as a sister taxon to *Heterocephalus* was more weakly supported in the 12S rRNA molecular phylogeny of Allard and Honeycutt (1992) than in the combined cytochrome-*b* and 12S rRNA molecular phylogeny of Faulkes *et al.* (1997a). In the latter, there is good support for placing *Heliophobius* and *Heterocephalus* as sister taxa, and as ancestral forms in the family. Such a grouping also makes sense in terms of the biogeography, as the family then splits into eastern-African and central/southern African clades, i.e. *Heliophobius/Heterocephalus* and *Bathyergus/Georychus/Cryptomys*. If this scenario is correct, Allard and Honeycutt (1992) suggest a radiation from eastern Africa, with species moving westwards and south into central/west Africa and southern Africa. This may have occurred by movement of animals through an 'arid corridor' between east Africa to southern Africa that has closed and reopened during climatic fluctuations since the Miocene (Van Couvering and Van Couvering, 1976; Honeycutt *et al.*, 1991a). The age and distribution of the bathyergid fossil record from the early Miocene, which includes specimens from the Namib and East Africa, lend some support to this hypothesis (Lavocat, 1978), but do not preclude the possibility of a central African ancestor radiating into east and southern Africa.

Chapter 2

The subterranean niche

2.1 BURROW ARCHITECTURE

The African mole-rats spend their entire lives underground, effectively almost completely sealed from the surface in an extensive labyrinth of foraging burrows and chambers (Nevo, 1979; Jarvis and Bennett, 1990). Bathyergids generally do not come above ground, although exceptions to this rule occur in some of the solitary species in the Cape Province. For example the Cape mole-rat is sometimes known to travel above ground before establishing new burrows at a different site (Jarvis and Bennett, 1991). Apart from these comparatively rare events, mole-rats remain underground whilst foraging for their food and when disposing of excavated soil onto the surface. With the exception of the naked mole-rat, which disposes of excavated soil from open 'volcano' shaped mounds, all other bathyergids retain a plug of soil between themselves and the molehill during the process of excavation. Naked mole-rats are particularly vulnerable to predation when flicking soil out of the burrows from their molehills (see Section 2.7). The retention of a soil plug by other species between the mole-rat and the surface reduces the risk of predation, but also means that the opportunity for gaseous exchange between the burrow and the outside world, as well as exposure to light, are minimal. This leads to a special and somewhat extreme set of conditions within burrows, which have given rise to specific physiological and behavioural adaptations in the various mole-rat species.

The burrows of all mole-rats have a similar architecture, consisting of a complex network of numerous superficial foraging galleries and deeper, more permanent, 'highways'. Amongst the latter part of the burrow is a multi-entranced nest chamber that may, or may not, have a closely associated food store where collected geophytes are stored, packed in soil (presumably enhancing their 'shelf life'). Single-entranced toilet chambers are also present in the burrow, but these are usually sited away from the nest (Jarvis and Bennett, 1991). In social species, the communal nest and toilet chambers are important focal points, possibly acting as 'information centres' where individuals can interact socially on a regular basis. The closed burrow system provides a buffered environment for the mole-rat, offering protection from environmental extremes as well as most predators. The burrow is extended mainly as the animals go in search of food (see Chapter 3). Indeed, it is the profusion and diversity of the plant types upon which they feed that have enabled the Bathyergidae to radiate into the wide range of environments in which they are found today. Two genera of solitary mole-rats, *Bathyergus* and *Georychus*, supplement their diet with aerial vegetation such as grass and the flowers of annuals, which are drawn down into the burrow and consumed (Davies and Jarvis, 1986). Small geophytes (which are readily transportable) are either eaten *in situ* or moved from the point of discovery to a strategically placed central food store. Larger bulbs and tubers are normally fed upon *in situ*.

Most burrowing activity occurs during, or just after rains when the soil is softened and easily workable. All genera excavate the soil with the extra-buccal incisors, except *Bathyergus,* where well-developed claws are employed. In all species, the fore and hind limbs are used to transport the loosened soil behind them (Genelly, 1965; Jarvis and Sale, 1971; Bennett, 1988). The mole-rat moves backwards as the soil is shunted along the burrow until it reaches the point at which soil expulsion is to take place. The excavated soil is expelled onto the surface by the mole-rat's powerful hind legs to form a characteristic dome-shaped mound. Stiff hairs on the outer borders of the tail and hind feet facilitate in the collection and transportation of excavated soil.

Burrow length appears to be related to the availability of food in the territory, in addition to the number of individuals within the burrow system (Jarvis and Sale, 1971; Hickman, 1979; Jarvis, 1985; Davies and Jarvis, 1986; Brett, 1986). The burrow systems of the social mole-rats (the common, Natal and naked mole-rats) have a lower biomass per metre of burrow (0.8, 1.1, and 0.9 to 2.1 g m^{-1}

Table 2.1. A comparison of burrow lengths and biomass for the Bathyergidae. In species where sexual dimorphism occurs, mean adult body mass is given for both males (m) and females (f). Jarvis and Sale, 1971[1]; Davies and Jarvis, 1986[2]; Jarvis and Beviss-Challinor, unpubl.[3]; Taylor et al., 1985[4]; Hickman, 1979[5]; Jarvis, 1985[6]; Brett, 1991a[7] (modified from Davies and Jarvis, 1986).

Species	Mean adult mass (g)	No. moles per burrow	Mean biomass per burrow (g)	Mean burrow length (m)	Mean ± S.D. biomass/ burrow length (g m^{-1})	Approx. home range of burrow (m^2)	Habitat
Heliophobius argenteocinereus[1]	160	1	180	47	2.1 ± 0.3 (n=4)	172	Athi Plains, Kenya
Bathyergus suillus[2]	933 (m) 635 (f)	1	877	256	3.8 ± 1.8 (n=7)	1390–3496	Western Cape Fynbos
Georychus capensis[3,4]	181	1	169	47	4.5 ± 1.9 (n=14)	272	Western Cape Fynbos
Cryptomys h. hottentotus[2]	83 (m) 58 (f)	2–14	132–810	464	0.8 ± 0.2 (n=4)	3922	Western Cape Fynbos
Cryptomys h. natalensis[5]	106 (m) 88 (f)	2-3	202	181	1.1 ± 1.0 (n=4)	2550	Grasslands of Natal
Heterocephalus glaber[6]	21	60	1250	595	2.1 (n=1)	5401	Semi-desert, Lerata, northern Kenya
Heterocephalus glaber[7]	33	87	2650	3022	0.9 (n=1)	105000	Wooded grassland, Kamboyo, southern Kenya

respectively) than do the solitary mole-rats (the Cape dune and Cape mole-rats) at 3.8 and 4.5 g m^{-1} respectively (Table 2.1; Hickman, 1979; Davies and Jarvis, 1986; Brett, 1986; Jarvis and Bennett, 1990). Interestingly, the mean adult body mass for naked mole-rats was different for southern and northern Kenyan animals (Table 2.1), with the latter being around 39% smaller, although group sizes are broadly similar. This may result from differences in available food biomass, because large *Pyrenacantha kaurabassana* tubers are mainly absent in the Lerata region of northern Kenya (see Chapter 3). Thus, naked mole-rats may respond to reduced food availability not by reducing the workforce, but by having a lower mean body size within each colony (Jarvis, 1985).

Burrow length can be very variable as it is constantly being restructured. The blocking of old sections routinely takes place, in particular during the dry season when the effort involved in pushing mounds on to the surface is great. Although burrow length can vary, there is a trend for it to be shorter for the solitary species like the Cape mole-rat (J.U.M. Jarvis, unpubl.), than for social species like the naked mole-rat which may have extensive burrow systems exceeding 3 km (Brett, 1986, 1991a; Figure 2.1). Limited field data for the common mole-rat (C.M. Rosenthal, N.C. Bennett and A.C. Spinks, unpubl.) and the Damaraland mole-rat (Jarvis and Bennett, 1993, unpubl.) suggest that although the burrow pattern is constantly changing, the overall home range appears to remain remarkably consistent, possibly reflecting the higher cost of digging and the more patchy distribution of food in their habitat (see Chapter 8). Solitary species tend to be less faithful to an area and appear to get up and go! Nanni (1988) found that of six Cape mole-rats he recaptured after between seven and 14 months, three had moved between 109 m and 210 m from the site of first capture.

Digging statistics are impressive, especially for the relatively small naked mole-rats. By weighing molehills and estimating the corresponding tunnels that would have produced them, Brett (1991a) calculated that in one rainy season in 1983, a colony of 87 animals excavated 1 km of new tunnels. Between 1982 and 1983, the same colony excavated between 2.3 and 2.9 km of new tunnels. These figures are probably conservative estimates as they would not include soil shifted around within the burrow and not moved to the surface. Similarly impressive are the digging figures estimated for the Damaraland mole-rat, where a colony of 16 animals with a total biomass of 2.2 kg excavated and moved 2.6 tonnes of soil in less than two months.

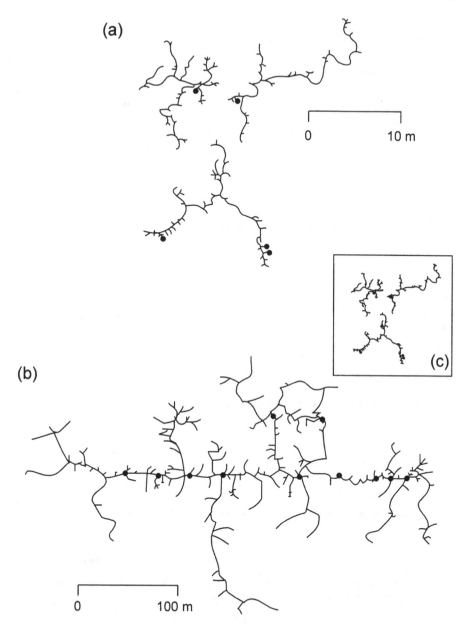

Figure 2.1. Plans of the burrow system of (a) three sub-adult *Heliophobius argenteocinereus* burrows, Athi Plains, Kenya (tunnel lengths of 45, 39 and 44 m) (adapted from Jarvis and Sale, 1971); (b) a naked mole-rat burrow in Mtito Andei, Kenya (total tunnel length 3022 m) (adapted from Brett, 1991a); (c) the burrows shown in (a) drawn to the same scale as the naked mole-rat burrow in (b), for direct size comparison. Filled circles denote the locations of nest chambers.

(a)

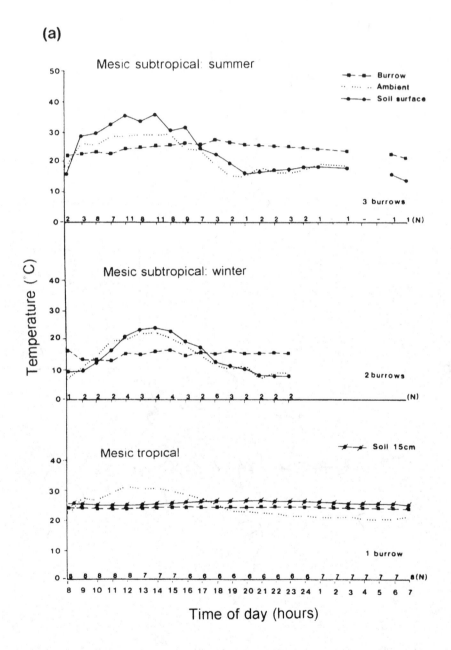

Figure 2.2. Daily changes in mean ambient, soil surface and burrow temperatures in (a) mesic, and (b) xeric areas (from Bennett *et al.*, 1988).

(b)

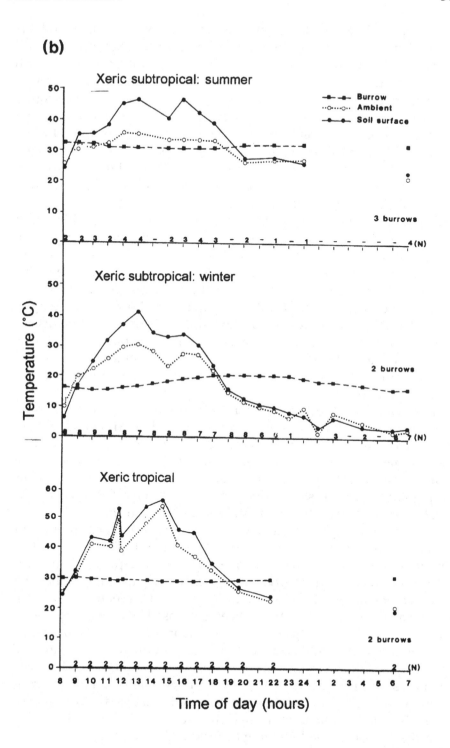

Owing to their small size and thermal sensitivity, many rodents have taken to a subterranean existence and utilise burrows to avoid hostile environments (Schmidt-Nielsen, 1975). Inhabitants of such sealed systems are consequently exposed to very different environmental conditions compared with other above-ground terrestrial mammals. Indeed, physical conditions such as light intensity, humidity, temperature and gas composition deviate considerably from those values normally experienced above ground (Schmidt-Nielsen *et al.*, 1970; Baudinette, 1972). The burrow provides the subterranean animal with, on the one hand, a relatively safe, thermostable niche, but, on the other, a potentially hostile environment that is characterised by high relative humidities, high carbon dioxide levels and low oxygen concentrations (Kennerly, 1964; Darden, 1972). The most extensive portion of the burrow system of mole-rats is superficial and lies 15 to 20 cm below the surface of the ground. Although showing some diurnal and seasonal fluctuations in temperature (Bennett *et al.*, 1988), superficial burrow temperature fluctuations are far more muted than those occurring at the surface. The sealed burrow shares a similar overall daily pattern of change with respect to soil surface and air, but the magnitude of the changes and the times of day at which maximal and minimal temperatures occur are different.

The lowest soil surface temperatures occur at, or just prior to, sunrise, and temperatures peak around mid-afternoon (15:00 to 16:00). The lowest burrow temperatures occur around 10:00 to 11:00 and the highest well after sunset (Figure 2.2). The daily reversal of heat flow is such that twice every 24 hours soil surface and burrow temperatures are the same (Bennett *et al.*, 1988). The time of the 'double cross' (Kennerly, 1964) can be seen in Figure 2.2.

The greatest range in annual and daily subsoil temperatures occur within the first 30 cm of soil (Bennett *et al.*, 1988). This may explain why mole-rats excavate nest chambers at greater depths, where there is greater thermal stability. The depth of the nest chamber characteristically exceeds 40 cm and in the Damaraland mole-rat it may be as deep as 2.4 m! It has been proposed that the deeper nest serves as a refuge from predators. The main predators of mole-rats are snakes, which rarely venture deep into the burrow system, instead preferring to take their prey from more superficial tunnels. However, it would seem more likely that the nest serves a thermoregulatory role, whereby mole-rats can retreat to a thermally stable depth, buffered from the daily fluctuations which are exhibited in the more superficially constructed foraging galleries.

2.2 PHYSICAL AND SENSORY ADAPTATIONS TO LIFE UNDERGROUND

The African mole-rats are totally adapted to a subterranean existence. In fact the name Bathyergidae is derived from the Greek *bathys*, meaning deep, and *ergo*, meaning to work. Mole-rats are characterised by cylindrical bodies and relatively short legs, giving the animal a low body carriage. This streamlined form allows mole-rats to move backwards as easily as they move forward. The tail is short (except in the naked mole-rat), fringed by a brush of coarse hairs and is used in moving soil along burrows. The head in all species is blunt and strong, with a muscular but indistinct neck. The eyes are small, with a thickened cornea, and there are no external ear pinnae, presumably eliminating the chance of abrasion whilst moving down the tunnels of the burrow. Other notable features of the mole-rat are the pig-like nose and the two pairs of large, ever-growing incisor teeth that are actually situated outside the mouth (extra-buccal) to prevent ingress of soil during digging. A number of fine vibrissae stand out above the body pelage and are used in much the same fashion as the facial whiskers of other rodents. These vibrissae are even present in the naked mole-rat, which lacks a normal pelage entirely (Thigpen, 1940). The importance of information conveyed via the whiskers is well documented in rats and mice, where each individual facial whisker has its own barrel field, a discrete area in the cortex of the brain dedicated to processing incoming stimuli (Woolsey *et al.*, 1975). These barrel fields have also been reported in a subterranean mammal, the star-nosed mole, *Condylura crista* (Catania and Kaas, 1995). Sensory vibrissae may function to enable mole-rats to orientate themselves when travelling down the narrow tunnels of their burrow, and to increase their sensitivity to other tactile cues.

The subterranean environment clearly imposes constraints on the mode of communication among individuals. In common with other subterranean animals, the visual system of the naked mole-rat is degenerate compared with terrestrial diurnal mammals (Eloff, 1958), and social information is transmitted via other sensory modalities, i.e. by the use of tactile, olfactory and auditory cues, although even airborne sound may be restricted in small tunnels (Narins *et al.*, 1992; Heffner and Heffner, 1993).

In terms of olfaction, one would expect mole-rats to have a well-developed sense of smell, given their poor visual sense. Although little research has focused on the anatomy and physiology of the olfactory sense in mole-rats, naked mole-rats, at least, are known to possess a vomeronasal organ (D.H. Abbott, pers. comm.). This accessory olfactory system is specifically involved in the chemoreception of pheromonal signals (Harrison, 1987). From studies in captivity, naked mole-rats are known to use olfactory cues in colony recognition (Faulkes, 1990; O'Riain and Jarvis, 1997) and recruiting colony mates to food sources (Judd and Sherman, 1996).

The underground niche also has an influence on the auditory environment. The hearing capabilities of the naked mole-rat are degenerate compared with terrestrial rodents, and restricted in their range. They have a limited ability to detect sound, limited high frequency hearing and have difficulty in localising sound, compared with surface-dwelling rodents (Heffner and Heffner, 1993). Despite this, naked mole-rats have the largest known vocal repertoire of any rodent. Eighteen vocalisations are known, most of which are associated with particular contexts or behaviours (Pepper *et al.*, 1991). Recently, 13 vocalisations were characterised for a species of *Cryptomys* from Zambia (Credner *et al.*, 1997), suggesting that, at least in the social bathyergids, vocal communication has reached relatively sophisticated levels for a rodent.

Another interesting morphological feature possessed by several species of bathyergid mole-rats is a characteristic white head-patch or 'bles'. This feature is present in all animals in some species, e.g. the Damaraland mole-rat, usually absent in others, e.g. the common mole-rat, and in some species variable in its size, e.g. the Cape mole-rat and the Angolan mole-rat. The exact function of the bles is unknown, although it has been suggested that it may have functioned as a signalling mechanism in the evolutionary past (N.C. Bennett, unpubl.), or as a photic window for increased light reception to the pineal gland (Lovegrove *et al.*, 1993).

The absence of any appreciable exposure to light in the bathyergid mole-rats may not only affect their biological rhythms, but also their metabolism of certain nutrients. The formation of vitamin D_3, which is involved in the uptake of calcium from the gut, is dependent on ultraviolet light. In addition, there is no obvious source of dietary D_3 as mole-rats are herbivorous, and naked mole-rats are known to be naturally deplete in D_3 (Buffenstein and Yahav, 1991a). Adaptations in the naked mole-rat, the common mole-rat and the Damaraland mole-rat permit uptake of calcium from the gut that is independent of the vitamin D mediation seen in other mammals, and is extremely efficient. More than 90% can be absorbed, compared with the 60% efficiency of most mammals (Pitcher and Buffenstein, 1994, 1995). One consequence of this increased efficiency is that in captivity, naked mole-rats are prone to develop calcified internal organs and other lesions if given excessive dietary supplements (C.G. Faulkes, unpubl.; Buffenstein *et al.*, 1995).

2.3 PHYSIOLOGICAL ADAPTATIONS TO THE SUBTERRANEAN NICHE

In the confines of the nest, the mole-rats are probably exposed to high carbon dioxide (hypercapnic) and low oxygen concentrations (hypoxic), but at the same time their metabolic rates will be at their lowest. Adaptations to these hypoxic conditions are apparent. Blood respiratory properties of naked mole-rats have been investigated and, compared with mice, they have haemoglobin with a very high affinity for oxygen. This property of naked mole-rat haemoglobin appears to be inherent in the molecule, and is presumably an adaptation to the low oxygen environment (Johansen *et al.*, 1976). Whether other bathyergids also have such adaptations is unknown. A behavioural adaptation to reduce oxygen consumption and energy expenditure in naked mole-rats takes the form of huddling. This occurs when individuals are resting together in the nest chamber and, in experimental studies, compared to individual animals, huddled mole-rats were able to maintain a constant body temperature at a reduced metabolic cost, utilising less oxygen (Withers and Jarvis, 1980). At thermoneutrality (the temperature range within which metabolic heat production is unaffected by an ambient temperature change), most subterranean rodents have resting metabolic rates (RMRs) substantially lower than those of surface-dwelling rodents of a similar size. Generally, subterranean mammals are also characterised by low body temperatures and conductances that are high compared with those of surface-dwelling rodents whose body temperatures are higher and thermal conductances lower (Table 2.2). Bathyergid mole-rats are characteristic of subterranean mammals with low body temperatures, low RMRs and generally elevated thermal conductances (Table 2.2).

One theory proposed by McNab (1966) to explain these low RMRs and body temperatures and high thermal conductances is that they help to minimise overheating in the closed burrow system. Evaporative water loss and convective cooling normally employed to cool terrestrial mammals are only involved to a minor degree in the animal's heat balance because burrow humidities are high (McNab, 1966, 1979). It has also been suggested that a low RMR, along with a slow heart beat, may normalise the partial pressures of oxygen and carbon dioxide in the blood and tissues under the hypoxic and hypercapnic conditions characteristic of a sealed burrow system (Darden, 1972; Ar *et al.*, 1977; Arieli *et al.*, 1977; Arieli and Ar, 1981). There would certainly be a selective advantage to conserving oxidative energy and hence reducing metabolism in an environment with depressed oxygen and increased carbon dioxide concentrations. This is another good illustration of a physiological adaptation to life underground.

Table 2.2. Resting metabolic rate (RMR), body temperature and thermal conductance estimates for surface-dwelling and subterranean mammals, compared with the Bathyergidae. References as follows: [1]McNab, 1979; [2]Morrison and Ryser, 1952; [3]Lovegrove, 1986a,b; [4]Bennett *et al.*, 1994b; [5]Lovegrove, 1987; [6]Bennett *et al.*, 1992; [7]1993a,b; [8]Buffenstein and Yahav, 1991b.

	RMR $(cm^3 O_2 g^{-1} h^{-1})$	Body temperature (°C)	Thermal conductance $(cm^3 O_2 g^{-1} h^{-1} °C^{-1})$
Surface-dwelling mammals[1,2]	1.0–1.4	36–40	0.009–0.080
Subterranean mammals[1,3,4]	0.4–1.2	32–36	0.070–0.190
Bathyergidae[3,4,5,6,7,8]	0.4–1.0	32–35	0.050–0.190

Within the family, thermoregulatory abilities range from the naked mole-rat, which resembles a poikilothermic mammal, to those that are more typically homeothermic. The naked mole-rat has the poorest thermoregulatory ability in the family (McNab, 1979; Buffenstein and Yahav, 1991b). Buffenstein and Yahav (1991b) found that the naked mole-rat showed a complete lack of control of body temperature over the entire range of ambient temperatures to which they were subjected (10° to 38 °C). The findings were extremely unusual for a eutherian mammal, with body temperature being essentially proportional to ambient temperature, producing an isometric relationship. While these results suggested that the naked mole-rat is acting like a poikilotherm, these trends are not reflected in their tissue respiratory capacity, because their tissue oxygen uptake and enzyme activities were typical of warm-blooded homeotherms (Gesser *et al.*, 1977). Thus, it would appear that, while naked mole-rats do show some poikilothermic traits as a result of adaptation to a thermostable environment, their basic biochemistry and physiology are that of a homeotherm. The inability of the naked mole-rat to maintain a near constant body temperature is due to their high surface area to volume ratio and limited insulative covering, giving a high thermal conductance. Although they lack fur, they do possess a thin layer of subcutaneous fat. An alternative suggestion as to why they may exhibit a form of poikilothermy could be an inability to produce endogenous heat to control body temperature (Buffenstein and Yahav, 1991b). Below 29 °C, oxygen consumption rose concomitantly with increasing body temperature (a typically poikilothermic trait). Above 29 °C oxygen consumption follows a typical endothermic pattern. The thermoneutral zone for naked mole-rats lies at a temperature normally encountered in the burrow.

Another rather bizarre thermoregulatory characteristic of naked mole-rats has been observed by Yahav and Buffenstein (1992). In mole-rats, the high-fibre diet necessitates fermentation in the hind-gut (caecum) by microbial organisms to aid digestion. These fermentation reactions are normally exergonic, producing heat (as, for example in ruminants) but, surprisingly, Yahav and Buffenstein found that in the naked mole-rat, the temperature of the caecum was actually *lower* than its abdominal surroundings. How this occurs is not clear, as there is no apparent anatomical vascular feature that might aid in the dissipation of heat, but again this abdominal 'heat sink' could be yet another adaptation to prevent overheating.

The Damaraland mole-rat and the common mole-rat both show a homeothermic (warm-blooded) relationship of body temperature (Tb) to ambient temperature (Ta), with little variation in body temperature over a wide range of ambient temperatures (Bennett *et al.*, 1992). However, while the Mashona mole-rat shows a more typically homeothermic profile with regard to its oxygen consumption and hence metabolic heat production, it exhibits heterothermy with respect to its body temperature. It has the lowest body temperature of the haired bathyergid mole-rats. Below 25 °C, the Mashona mole-rat experiences problems in maintaining a stable core temperature. Like the naked mole-rat, this may be attributed to its small body mass and increased surface area to volume ratio giving a high thermal conductance (Bennett *et al.*, 1993a). When in thermal equilibrium, Ta and Tb can normally be regulated by balancing heat loss with heat production. When the Mashona mole-rat is at temperatures below

25 °C, despite a marked increase in oxygen consumption over that at thermoneutrality (up to 260% at 18 °C), the increase in heat production is insufficient to maintain a constant Tb over the range 25 to 18 °C. At temperatures below 18 °C, the Mashona mole-rat exhibits strong heterothermic tendencies. The wide range of thermoregulatory capabilities exhibited in the Bathyergidae illustrates the versatility of the family in adapting to differing thermoregulatory constraints in what is otherwise a remarkably uniform niche.

The burrows of mole-rats occurring in the latitudes away from the equator exhibit marked seasonal changes both in mean burrow and above-ground temperatures, whilst little seasonality occurs in the tropics (Bennett *et al.*, 1988). However, the seasonal component in the burrows of mole-rats inhabiting higher latitudes is still considerably muted when compared with the seasonal changes occurring above ground. It is plausible that these regional differences in temperature profiles may lead to the seasonality of breeding as shown in mole-rats of higher latitudes (the common, Cape and Cape dune mole-rats) and the absence of seasonality in more equatorial species (the naked mole-rat). In most terrestrial mammals, changes in photoperiod over the course of the year and over daily periods act as the main environmental cue that modulates both seasonality in breeding and daily (circadian) patterns of behaviour. In the mole-rat's environment this photoperiodic cue is absent, so, since their visual system is degenerate, other environmental factors like temperature and rainfall probably have an overriding influence.

2.4 ACTIVITY PATTERNS IN AFRICAN MOLE-RATS

Perhaps, as a result of the lack of photoperiodic cues, most subterranean rodents exhibit an evenly distributed activity pattern. Laboratory studies conducted upon the pocket gopher, *Geomys busarius* (Vaughan and Hansen, 1961) and radiotelemetry studies on free-ranging pocket gophers, *Thomomys talpoides* (Andersen and MacMahon, 1981) and *T. bottae* (Gettinger, 1984) suggest activity is evenly distributed over the 24-hour cycle. In the bathyergids, field studies on the silvery mole-rat (Jarvis, 1973a) and laboratory studies on the Natal mole-rat (Hickman, 1980) and the common mole-rat (Bennett, 1992) have also revealed a dispersed activity pattern throughout the 24-hour period. In the Damaraland mole-rat, radiotracking of wild animals by Lovegrove (1988) revealed that activity patterns were not influenced by daily fluctuations in burrow temperature or photoperiod *per se*, but that activity was influenced by the risk of hyperthermia in the superficial foraging tunnels.

The nest chamber is the focal point of the burrow system where individuals of social species can come together on a regular basis and interact socially. In captivity and in the wild, mole-rats spend a large proportion of time resting and sleeping within the nest chambers. However, bathyergid mole-rats generally lack a distinct sleeping period. In the social mole-rats such as the Damaraland mole-rat there is asynchrony of each mole-rat's activity cycle with that of other colony members (Bennett, 1992; Kinloch, 1982). Hence, at any particular time, an active mole-rat is usually present within the burrow. In naked mole-rats studied in captivity, Davis-Walton and Sherman (1994) found no evidence for regular sleep–wake cycles, although there was a tendency for colony mates to arise and retire simultaneously.

Interestingly, the breeding female of the naked mole-rat was found to be the most active colony member. In the wild, the queen was found to be asleep during only 30 to 40% of sampled intervals, compared with others in the colony who slept for 45 to 70% of the time (Brett, 1986). Increased activity in the queen may reflect her need to patrol the burrow, maintaining dominance and inciting activity among workers (Davis-Walton and Sherman, 1994). Similar patterns have also been seen in laboratory studies of the common mole-rat (Kinloch, 1982; Bennett, 1992), where the reproductive animals were found to spend between 36 and 49% of the time in the nest compared to between 70 and 81% in non-reproductives. Similar extended periods of rest by non-breeders in the confines of the nest have been found in the Damaraland mole-rat both in the laboratory (Bennett, 1990) and in the field (Lovegrove, 1987; J.U.M. Jarvis and N.C. Bennett, unpubl.).

2.5 BEHAVIOURAL ADAPTATIONS TO A SUBTERRANEAN LIFESTYLE

Apart from physiological and sensory adaptations, the subterranean niche also requires behavioural specialisations for foraging, maintenance and upkeep of the burrow. Four main behaviours fall into the category of burrow maintenance. These are digging and gnawing, sweeping and transport of soil in the burrow, nest building and food carrying.

In the wild, nest building and the carrying of nesting material involve transporting husks which have been removed from tubers and root epidermis from harvested geophytes. These are the only materials which have been found in nests in the field. In captivity wood-wool and paper towelling provided the source of nesting.

The social species of mole-rat share a common cache of food items (see Chapter 3). In the wild, corms, bulbs and tubers and other small food items are carried to this communal food store between the animal's incisors with the head held high. In the common mole-rat, Bennett (1992) found that juveniles carry food significantly more frequently than adults, whilst in colonies of the Damaraland mole-rat, it was the smaller mole-rats (frequent workers) that carried food to the store more regularly than the larger infrequent workers (Bennett, 1990). In the naked mole-rat, there is a significant negative correlation between body mass and carrying food (Jarvis, 1991; Lacey and Sherman, 1991; Chapter 4).

Figure 2.3. Sequence of burrowing behaviour. The chisel-like extrabuccal teeth are used to displace soil from the work face. The accumulating soil is moved along the tunnel by scraping it under the abdomen and kicking out behind the animal.

Burrowing in the bathyergids has been described by Genelly (1965) and Jarvis and Sale (1971). It involves an initial excavating period during which the extremely sharp extrabuccal incisors are used to dig away at the substratum. The main digging thrust against the substratum is with the lower incisors. During digging the animal braces itself with the hind feet, whilst its forefeet are used to gather and scrape the excavated soil under its abdomen. The hind feet are then brought forward and they kick the soil behind them. In this way a pile of soil accumulates behind, which the animal periodically reverses onto and pushes down the burrow using the hind feet. The forefeet provide the backward thrust, whilst the nose and incisors are pressed against the burrow floor or roof, providing additional leverage. The journey along the burrow consists of a series of abrupt movements in which the forefeet are initially close to the hind feet, and the back is arched. This is followed by a powerful straightening of the back and the extension of the fore and hind limbs. Periodically, the mole-rat may stop to gather any loose soil that has been missed (Figure 2.3). Naked mole-rats may also form characteristic 'digging chains', where groups of animals form a line, with the one in front excavating the soil, which is then swept backwards by each animal in turn, rather like a conveyer-belt. Periodically, the digging animal will be physically dragged back and replaced by the next in line (Lacey and Sherman, 1991).

As in other rodents, the incisors of the bathyergids grow continuously. In the wild, the incisors are worn down as they dig by chiselling through the soil with their teeth, and therefore tooth wear is potentially an important constraint. In captive animals the teeth are honed by tooth-sharpening, a behaviour observed in all bathyergids studied to date (Jarvis, 1969b; Bennett, 1988, 1990, 1992). The process of tooth-sharpening is essentially the same in all the mole-rats and usually occurs in the nest area when the animal is drowsy, often preceding sleep. The mole-rat initially braces itself with its forepaws whilst adopting the tooth-sharpening posture. Forward and backward and side-to-side actions of the lower jaw are then used to file the upper and lower incisors against one another. Short forward thrusts of the lower incisors against the upper ones sharpen the lower incisors, whereas longer and slower backward movements of the lower incisors across the upper incisors sharpen the upper. Careful examination of this process reveals that flakes of chipped incisor are periodically flicked out of the mouth by the tongue (Figure 2.4).

It has been shown by Bennett (1990) that, in the Damaraland mole-rat, 'frequent workers' sharpen their teeth significantly more frequently than 'infrequent' workers. It is possible either that the 'frequent workers' have a greater incisor growth compared with the rest of the colony, or that their activities blunt these 'digging tools' at a faster rate. This latter suggestion is supported by work by Brett (1986) who noted that when naked mole-rats dug through very hard soils they had to make frequent stops to sharpen their incisors.

Characteristic behaviours also occur in the toilet chamber, a well-defined area within the burrow system frequented by all mole-rats, and usually a blind-ending tunnel section. After the nest, this is the second most important focal area in the burrow. There are essentially three types of toilet behaviour: grooming and smearing, urination and defecation.

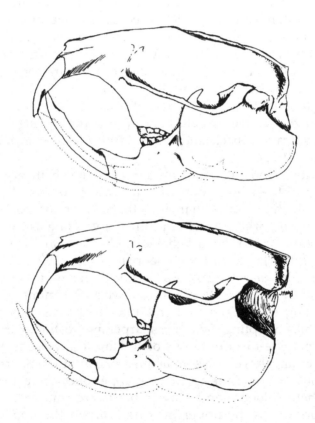

Figure 2.4. Incisor sharpening is an important behaviour in that it keeps the 'digging tools' of the mole-rat well-honed. The upper and lower incisors are moved relative to one another in order to keep them sharp, and at the optimum length.

In grooming and smearing, the mole-rat enters the toilet area and spends some time smelling the area and then vigorously grooming its head region, flanks and belly. As it leaves the toilet area, it drags its ano-genital region along the burrow, leaving a small trail of fluid along the tunnel. The animal drags its rear for about 30 to 45 cm before adopting its natural posture. The exact function of grooming in the toilet area and the subsequent smearing of a fluid on exit from the latrine is unknown, but it appears to occur mainly within the social bathyergids (Bennett, 1988). Smear marking may familiarise each mole-rat with the odour of other colony members. This would enable members of the colony to differentiate between conspecifics and intruding individuals. Marking may be of importance in expressing the internal sexual status of each animal, the urine reflecting the plasma hormone concentration.

2.6 ENERGETICS AND THE EFFECT OF SOIL HARDNESS ON BURROWING BEHAVIOUR

High energetic costs are incurred as subterranean mammals excavate their burrow systems and transport the loosened soil to the surface. Depending on soil hardness and burrow diameter, the cost of burrowing has been estimated to be 360 to 3600 times that of travelling the same distance on the surface (Vleck, 1979, 1981). As we have mentioned, all genera of bathyergid mole-rats excavate soil using their extra-buccal incisors, and the two species of *Bathyergus* use the foreclaws as well. The rate of wear on the chisel-like extra-buccal incisors and the rate at which these ever-growing 'digging tools' are replaced are therefore additional constraining factors on the activity of mole-rats.

The nature of the substrate is also important in burrow excavation, and in particular the water content of the soil can have important consequences on burrowing. In the arid regions inhabited by the naked and Damaraland mole-rats, the sandy soil can be unworkable for much of the year (Leistner, 1967), as the soils can be either extremely hard to burrow through (e.g. in certain localities of the naked mole-rat habitat in East Africa), or the surface 40 cm of soil may be extremely soft and fluid (e.g. in the sand dunes of the Kalahari, a habitat common to the Damaraland mole-rat). Stony soils tend to be avoided by the mole-rats. Hard soils have an abrading effect and incisor wear is greater in these soils. On the other hand, the soft sands of the dunes are easy to dig, but the loose soil cannot be readily compacted since its fluid nature results in subsidence of soil into the region of excavation. It becomes readily apparent that the best time to excavate and search for the food resource is during the wet season, when the soils are moist and easily worked and the energetic cost of burrowing is reduced. Hill *et al.* (1957) and Jarvis and Sale (1971) noted that naked mole-rat molehills seemed to be associated with rain showers. They suggested that this occurred because the efficiency of burrowing increases with a softening of the soil, which is normally baked like clay. Brett (1986) found from digging trials in the field that naked mole-rats can burrow more efficiently and excavate more effectively in moist soils than in hard soils. These findings were corroborated by Lovegrove (1989) who investigated the energetics of burrowing in both types of soil conditions. Using Vleck's (1979) model to predict burrowing costs, Lovegrove (1989) found that the cost of burrowing as a function of distance burrowed in two arid-dwelling mole-rats species is, overall, lower in damp soils than in dry soils. This means that per unit energy expended, mole-rats take a shorter time to dig a unit length of tunnel in damp soil, compared to dry soil.

However, if looked at in terms of instantaneous measurements of metabolic rate, burrowing in damp soil is actually energetically slightly more costly than the same behaviour in dry soils. Lovegrove (1989), using trials in captivity, showed that in tunnels packed with damp soil, the Damaraland mole-rat had a burrowing metabolic rate of 2.86 cm^3 O$_2$ g^{-1} h^{-1} whilst in dry sand it was 2.58 cm^3 O$_2$ g^{-1} h^{-1}. A similar pattern was found for the naked mole-rat, where in damp soil the cost of burrowing was 3.36 compared with 2.78 cm^{3}O$_2$g^{-1}h^{-1} in dry soil. These results initially seem to be a direct contradiction of the idea that costs are reduced when burrowing in damp soil. However, mole-rats appear to be able to dig more efficiently and with greater speed in the more manageable moist soil. This expains why the Damaraland mole-rat and naked mole-rat engage in extensive burrowing in their respective habitats immediately post rain. Apart from avoiding the physical problems involved in constructing molehills, in damp sand the Damaraland mole-rat utilises 3.7 times less energy overall when extending the 'exploratory' or primary burrow than would be the case in dry sand.

Interestingly, if one compares the ratio of digging metabolic rate to resting metabolic rate in damp sand, it is found that the three mole-rats, *H. glaber*, *C. damarensis* and *G. capensis*, have a fairly uniform ratio of around 5 (Table 2.3). This ratio is higher than that derived for the geomyid pocket gophers, whose values range from 2.5 to 4.8. This may be due to the fact that gophers are fossorial, and push out soil at burrow entrances, in contrast to mole-rats which are totally subterranean and throw mounds onto the surface without ever coming above ground. The higher cost of excavating in dry soils would in essence select for 'refined' harvesting of areas previously excavated during the short periods of extensive foraging. Jarvis *et al.* (1998) have shown that previously excavated areas which are rich in geophytes are often reworked once dry conditions set in. This 'farming' practice ensures that during the more extreme times of the year when excavation is more costly and energetically undesirable, mole-rats minimise their energy expenditure for obtaining geophytes. When excavating their tunnels all species, except the naked mole-rat, push up characteristic cores of soil which in turn dry and crumble to produce the characteristic dome-shaped mound of the molehill. These structures can be used by the trained eye as indicators of the whereabouts of inhabited burrows, and the approximate time since the molehill was produced. Molehills can have two distinctive forms. The first type are those which are associated with the formation of primary exploratory tunnels. The second type of molehill is usually associated with feeding and harvesting. These are typically either a single mound larger than the general mound associated with primary exploratory tunnelling, or two or three mounds which have been pushed up in close proximity to one another (Lovegrove and Painting, 1987).

Table 2.3. Resting (RMR) and digging metabolic rate (DMR), for selected subterranean rodents (modified from Lovegrove, 1989, © University of Chicago Press).

Species	Family	Mean body mass (g)	RMR ($cm^3 O_2$ $g^{-1} h^{-1}$)	DMR ($cm^3 O_2$ $g^{-1} h^{-1}$)	DMR/ RMR
Heterocephalus glaber	Bathyergidae	32	0.64	3.36	5.25
Cryptomys damarensis	Bathyergidae	152	0.57	2.86	5.02
Georychus capensis	Bathyergidae	197	0.59	3.41	5.78
Thomomys talpoides	Geomyidae	75	1.65	4.08	2.47
Thomomys bottae	Geomyidae	143	0.84	4.10	4.88

In the Damaraland mole-rat, excavation of lateral tunnels to deposit soil on the surface during the dry season is hampered, as we have said, by loose sand flowing back down into the burrow. This is because there is no compaction of the soil due to the soil moisture content being low. As a result there is little or no mound production in the dry season. The first good rains (providing they penetrate to an adequate depth) herald the first signs of new molehill production. The post-rain burrowing activity is associated with the extension of numerous near linear primary burrows, into an as yet unharvested region of geophytes. The excavation of these long linear primary burrows is an important strategy since it optimises the probability of finding geophyte-rich areas during times when efficient digging is possible. These can then be returned to and re-worked at a later stage, when secondary burrows are formed to optimally harvest the food resources.

In the dry season, while there is a conspicuous absence of molehills, activity is still occurring below ground. Instead of soil being moved to the surface, it is pushed around the burrow system and deposited in older, non-utilised sections. Although this behaviour may be energetically very expensive, it does enable the mole-rats to continue their subterranean foraging during periods when the low moisture content of the arenosols in the shallow sub-surface layer precludes extensive molehill production (Jarvis *et al.*, 1998).

Animals that burrow to find food pay heavily in terms of the energy they expend in finding it. Indeed, the energetics of finding widely dispersed and rare resources may have enforced group living in some of the species of mole-rats, and also selected for a low resting metabolic rate (see Chapter 8). Vleck (1981) suggested that in geomyid rodents (i.e. pocket gophers), body size is inversely proportional to soil hardness, since smaller body size requires a reduced burrow diameter and therefore less energy is expended in making tunnels. Lovegrove (1989) extended this finding to try to explain why a division of labour is observed in some of the social species of bathyergid mole-rats, whereby the naked mole-rat and Damaraland mole-rat (both occurring in harsh, arid environments) characteristically have workers with a small body mass. This observation suggested that there may have been selective pressures for a smaller body size in workers, when compared with the significantly larger infrequent workers and reproductives (see Chapter 4). Lovegrove (1989) calculated that if Damaraland mole-rats spent 30% of the day burrowing, and instead of 20 mole-rats with a mean body mass of 150 g burrowing, 50% of the colony maintained a mean body mass of 100 g, then this would increase the foraging efficiency of the colony. The reduction in mean body mass does not actually increase the foraging efficiency directly, but reduces the individual energy requirements. The daily energy expenditure of a 100 g Damaraland mole-rat would be about 32% lower than that for a mole-rat of 150 g, thus representing a daily saving in energy of about 16% for a colony of 20 mole-rats (Lovegrove, 1989).

The naked mole-rat may also reduce its colony energy budget by maintaining a small body size and a low resting metabolic rate. The Damaraland mole-rat, on the other hand, maintains the minimum body size and resting metabolic rate necessary for endothermy during the winter (Lovegrove, 1986a,b) and consequently incurs a slightly higher cost for burrowing (Lovegrove, 1989). So why do mole-rats show a frenzy of burrowing activity immediately after the rains? Apart from avoiding the physical problems involved in constructing a molehill in dry sand (Jarvis *et al.*, 1998), they also utilise less energy to extend the primary burrow when the sand is damp. Obviously, the further and faster a subterranean mammal or group of mammals burrows per day, the greater will be the average daily resource rewards. Following from this, the energetic rewards obtained when burrowing in damp soil should be greater, since the cost of burrowing has been estimated as 1.7 to 3.5 times lower than in dry soils (Lovegrove, 1989). Therefore, where soils are damp for only a short period of time, it would be energetically beneficial to a colony of mole-rats to extend their primary foraging burrows as rapidly and as far as possible, prior to the soil drying out and becoming unworkable again. The exploitation of resources in these initial excavations can then proceed more or less independently of the weather during the dry periods of the year, as soil can be moved sub-surface from one burrow section to another, without having to be voided onto the surface.

2.7 PREDATION

Although anti-predator behaviour is an important aspect of sociality in terrestrial cooperative breeders, relatively little is known quantitatively about the predation risk for mole-rats. Their subterranean environment certainly protects them from many of the typical above-ground predators. However, solitary species of mole-rats are still vulnerable when they are dispersing on the surface or when they come close to the surface whilst foraging for aerial vegetation. All species are at risk during molehill formation, particularly naked mole-rats which kick soil from the burrow (volcanoing) rather than retaining a plug of soil between the working mole and the surface (Jarvis and Bennett, 1991; Section 2.1). Mole snakes (*Pseudapsis cana*), a species sympatric with most, if not all, of the African mole-rats, appear to be able to detect such freshly turned soil and usually seize the mole-rat from behind before constricting and killing it. Both the mole snake and the cobra (*Naja naja*) have also been observed entering Damaraland mole-rat systems through parts of the burrow that have caved in (N.C. Bennett, unpubl. data). Brett (1991b) has witnessed the rufous beaked snake (*Rhamphiophis oxyrhynchus rostratus*), the file snake (*Mehelya capensis savorgnani*) and the white-lipped snake (*Crotaphopeltis hotamboela*) catching naked mole-rats as they kick soil from their burrow. In addition to snakes, mole-rats may also be preyed upon by some bird species such as owls and other birds of prey. Species with long, sharp beaks (e.g. storks and herons) could easily catch naked mole-rats during 'volcanoing', and Jarvis and Bennett (1991) have witnessed such an event with the common mole-rat. Mammalian carnivores such as caracal, jackal, hyena, viverrids and foxes may also capture mole-rats when they are close to the surface (De Graaff, 1981). Whether rates of predation differ in different habitats and species of mole-rats remains unknown, as very little fieldwork has tackled these issues.

Brett (1991b) noted that naked mole-rats may have adapted their burrowing activity and molehill production so that most occurs during the early morning and late evening when snakes are least active (due to torpor). This is potentially very important as snakes are also most active in the rainy season, when molehill production is at its peak. All predation events recorded by Brett (1991b) occurred during the rainy season.

It is readily apparent, even from the relatively limited studies reported here, that the bathyergid mole-rats are highly adapted to life underground, with many morphological and physiological specialisations. In the following chapters we will see that, while exploitation of the subterranean niche has definite advantages in terms of predator avoidance, reduced competition for food and the provision of a stable, predictable environment, a range of constraints must also be overcome. Principal among these are the high energetic costs of burrowing, which we have mentioned, and the risks of locating food by foraging (digging) blindly. The nature and distribution of the food resource will now be considered in Chapter 3.

Chapter 3

The food resource of African mole-rats

3.1 THE DIET AND ITS NUTRITIONAL CONTENT

Mole-rats are herbivorous, feeding upon the underground storage organs of geophytes, plants which possess perennating buds below the soil surface on a corm, bulb, tuber or rhizome. Mole-rats in the genera *Georychus* and *Bathyergus* also eat the aerial portions of grasses and forbs (Beviss-Challinor, 1980; Broll, 1981; Davies and Jarvis, 1986; Bennett, 1988). Geophytes are an ideal food resource for which there is apparently little competition with other species. They are also available for much of the year (when not flowering) and contain a high concentration of nutrients. Reabsorption of nutrients from senescing leaves and stems at the end of the growing season can be as high as 88% for phosphorus and 79% for nitrogen (Dixon, 1981). Many geophytes are unpalatable and toxic to most animals and contain cardiac glycosides (Watt and Beyer-Brandwijk, 1962), whereas others possess thick tunics and spinous coverings (Lovegrove and Jarvis, 1986). However, neither of these 'lines of defence' afford protection from the mole-rats, which seem to be immune to the toxins and whose chisel-like incisors readily penetrate the protective outer coverings.

The gut of all the bathyergid species is adapted to a high-fibre herbivorous diet, with the adults of all species exhibiting autocoprophagy. The hind gut/caecum of both the naked mole-rat, *Heterocephalus glaber*, and the Damaraland mole-rat, *Cryptomys damarensis*, contain large numbers of endosymbionts including

53

protozoa, bacteria and fungi (Jarvis and Bennett, 1991). On the whole, therefore, mole-rats appear to be strictly vegetarian, although there are reports in the literature of mole-rats occasionally having taken invertebrates which are commonly found in mole-rat burrows. Burda and Kawalika (1993) have found the giant Zambian mole-rat, *C. mechowi*, to consume earthworms and scarab larvae, whereas in the naked mole-rat, P.W. Sherman (pers. comm.) has found the head capsules of termites in the faeces of freshly captured animals. Jarvis (1991) has also reported that dead pups are eaten. While Burda and Kawalika (1993) have found the Zambian mole-rat to consume laboratory mice, the dentition of the mole-rats is suited to a herbivorous diet (Roberts, 1951) and omnivory is probably a rare event which takes place when the opportunity arises. Other studies that have examined the stomach contents of a number of freshly collected species of mole-rat, including the Cape dune mole-rat, *Georychus capensis*, the common mole-rat, *C. h. hottentotus*, the Mashona mole-rat, *C. darlingi*, and the naked mole-rat have revealed an absence of invertebrate matter in their gastro-intestinal tracts (Beviss-Challinor, 1980; Broll, 1981; Davies and Jarvis, 1986; Bennett, 1988; N.C. Bennett, pers. obs.). Brett (1991a) also found no evidence that mole-rats in his study area in Kenya ate any of the invertebrate fauna found in their burrows. However, he did find in one colony a blind-ending tunnel that emerged directly underneath a large bone lying on the surface (possibly from a zebra) and, during the period of observation, recorded three animals gnawing it. On examination, the bone was found to have been substantially chewed at one end. Brett (1991a) suggested that bones may possibly be an important supplement to the diet, providing calcium and phosphorus salts which are rare in the tuberous geophytes eaten by naked mole-rats. The teeth of the naked mole-rat wear rapidly during chisel tooth digging, but they also grow rapidly. Animal bones could therefore provide a source of minerals as substrates for tooth growth.

The geophytes encountered by mole-rats may be stored in carefully excavated food chambers and soil-packed blind endings of burrow branches (De Graaff, 1964, 1981; Genelly, 1965; Du Toit *et al.*, 1985; Davies and Jarvis, 1986; Lovegrove and Jarvis, 1986; Bennett, 1988). The corms and tubers in the stores are tended by the mole-rats which prune off, but do not eat, their buds and sprouting shoots. Evidence of disbudding has been found in the food stores of the Cape dune mole-rat, the common mole-rat and the Damaraland mole-rat (Davies and Jarvis, 1986; J.U.M. Jarvis and B.G. Lovegrove, pers. comm.). The storage and subsequent husbandry of geophytes is not unique to the family Bathyergidae. It has been reported in the Mediterranean spalacid mole-rats who store *Oxalis* bulbs. *Spalax ehrenbergi* differs from the bathyergids in that it eats the shoots from the apical region and does not discard them (Galil, 1967).

Jarvis and Sale (1971), Bennett (1988) and Spinks (1998) have all reported incidences of 'geophyte farming', whereby large geophytes are left *in situ* but the mole-rats partially consume them and then replug the remaining hollowed-out tuber with soil, thereby enabling the geophyte to continue growing. The farming of geophytes by the naked mole-rat has been suggested by Jarvis and Sale to circumvent the problems of desiccation and decomposition which could occur if small pieces of geophyte were excised and placed in the central store. Mole-rats adapt their foraging strategies to the size of the food resource. Tiny bulbs and corms are eaten as they are encountered, medium-sized bulbs (up to 4 cm) may be selectively cached, whereas large (>6 cm) bulbs and vertically extended larger tubers may be selectively 'farmed' *in situ.*

Geophytes typically possess a fibrous, or dry, outer husk which must be removed before feeding. The time involved in handling and preparing the geophyte is of importance with respect to time energy budgets. All mole-rat species studied to date completely remove the external husk from smaller geophytes before consuming them. Very large geophytes (>12 g) are rarely completely dehusked (Barnett, 1991; Spinks *et al.*, 1999b). The geophytes are processed in a very distinctive fashion. They are held between the forefeet, the bulb is chewed around the base and layers of the husk are peeled down' towards the tip, in much the same way as a banana is peeled (Figure 3.1). In many instances, the mole-rat first nibbles the tip of the shoot and samples the geophyte (Barnett, 1991). During dehusking, rapid vibratory movements of the forepaws are made whilst the bulb is held by the incisors. At this stage the choice is made as to whether to continue to feed or to discard the food item.

Mole-rats do not drink free water, but instead obtain all their fluid requirements from their food. Geophytes have a high moisture content at 77–80% by weight (Bennett and Jarvis, 1995). In the microclimate of the burrow, where humidity is very high, the loss of water from animals would be expected to be low (Buffenstein and Yahav, 1991b). In mole-rats, pulmonary and faecal water loss are generally not controlled (Urison and Buffenstein, 1994), although faecal water loss depends primarily on the fibre content of the diet; the higher the fibre content, the greater the loss of water (Buffenstein, 1985; Bennett and Jarvis, 1995). Because of the high moisture content of the

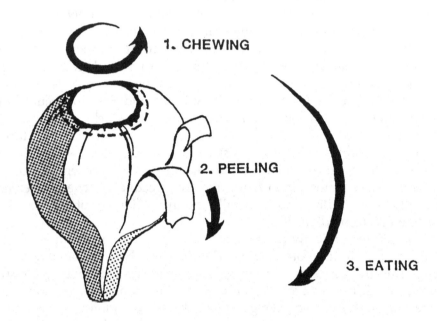

Figure 3.1. A schematic representation of how the outer tunic of a geophyte is de-husked (after Barnett, 1991).

geophytes, water conservation does not seem to have been a selective pressure. Naked mole-rats produce a fairly dilute urine and the amount of urinary creatinine, a protein metabolite excreted at a constant rate and therefore useful for estimating the concentration of urine, is generally low (Faulkes, 1990). Urison and Buffenstein (1994) found that the naked mole-rat had a moderate renal concentrating ability and could not maintain a plasma osmolality or body mass with either an extreme water stress or salt loading. In this species, it is suggested that successful habitation in arid regions is not facilitated by renal water conservation, but rather by underground existence in a microhabitat where there is high humidity and the radiant heat loads are low. This is probably true of all other species of mole-rats occurring in both mesic and arid habitats.

Foraging underground imposes high energetic demands on the mole-rats and they exhibit dietary and physiological modifications which enable them to exploit the resource to the full. Buffenstein and Yahav (1994) have shown that gemsbok cucumbers have a very low protein and high fibre content, which is similar to the nutritional content of poor-quality grasses that can usually be consumed by specialised ruminants. Despite this poor-quality food resource, the Damaraland mole-rat derives sufficient energy from the food by caecal fermentation. Lovegrove and Knight-Eloff (1988) found the caecum of the Damaraland mole-rat to comprise 26% of the total hind gut length, which is 3.7 to 20.2% larger than that currently recorded for any southern African rodent (Perrin and Curtis, 1980). The habit of coprophagy also enables the mole-rat to maximise the energy released from the fibre component by secondary fermentation.

All species of mole-rat, including those supplementing their diet with grasses and forbs, in addition to tubers, have similar digestive capabilities to one another with coefficients of digestion (> 90%) when fed on a uniform diet of sweet potato (Table 3.1). Hence, the differences in digestibility coefficients of food resources consumed by the various species of mole-rats are due to the food type and are not artefacts of differing digestibility abilities amongst the mole-rats.

In a survey of small mammals by Grodzinski and Wunder (1975), the highest digestibility and assimilation coefficients were found in the foods of granivorous rodents, where digestibility values reached 90% and above. At this time, there were no data available for small mammals feeding upon bulbs, corms and tubers. Since then, mole-rats consuming geophytes have been found to also have remarkably high coefficients of digestibility, which parallel that of seeds eaten by granivorous rodents (Bennett and Jarvis, 1995).

However, geophytes utilised by the mole-rats do not necessarily have high coefficients of digestibility. Markedly lower coefficients of digestibility were recorded for the fibrous gemsbok cucumber, where a digestibility of 50% was recorded for the Damaraland mole-rat fed exclusively on this diet. Interestingly, animals fed on this food item lost body mass (14% in seven days). This may mean that wild Damaraland mole-rats do not feed exclusively on the gemsbok cucumber or that they are highly selective of the age and portions of cucumber on which they feed. For example, two Damaraland mole-rats fed in the field for a couple of days upon young gemsbok cucumbers of lower fibre content had coefficients of digestibility around 70 to 80% (Bennett and Jarvis, 1995). Lovegrove and Knight-Eloff (1988) examined partially eaten tubers adjacent to mole-rat

Table 3.1. Food characteristics for natural and subtitute diets for five species of mole-rats, and the guinea pig (GE, mean gross energy ingested and egested; DE, digestible energy; S.D., standard deviation (after Bennett & Jarvis, 1995).

Species	Social status	Habitat	Food type	Energy content of food (kJ g⁻¹)	Mean GE ± S.D. in (kJ)	Mean GE ± S.D. out (kJ)	Mean % DE ± S.D.	Natural or substitute food
Georychus capensis	Solitary	mesic	sweet potato	15.5	156.6 ± 29.9	5.7 ± 2.8	95.7 ± 2.0	substitute
Bathyergus suillus	Solitary	mesic	couch grass stem and leaves;	15.9	362.0 ± 185	51.2 ± 15.6	83.2 ± 6.8	natural
			leaves only	15.4	535.6 ± 130	44 ± 5.1	90.8 ± 3.2	natural
Bathyergus janetta	Solitary	semi-arid	*Herria*	15.5	192.0 ± 24.3	34 ± 15	81.5 ± 7.2	natural
Cryptomys hottentotus hottentotus	Social	mesic/semi-arid	*Romulea* sp.; sweet potato	16.0	94.9 ± 14.5	3.2 ± 2.1	96.1 ± 1.8	natural
				14.7	274.5 ± 4.4	13.8 ± 4.4	94.1 ± 2.1	substitute
Cryptomys damarensis	Social	arid	sweet potato; gemsbok cucumber; *Dipcadi* sp.	15.5	124 ± 36	3.8 ± 1.8	96.3 ± 1.6	substitute
				14.2	87.9 ± 23	40 ± 12.5	53.0 ± 9.8	natural
				15.3	157.5 ± 19	6.3 ± 2.9	95.7 ± 2.2	natural
Guinea pig	–	mesic	couch grass leaves	15.3	329 ± 56.5	140.5 ± 17.2	55.7 ± 7.6	natural

Table 3.2. Mean water content, faecal water content, percentage ash, fibre and energy content of natural and substitute foodstuffs fed to mole-rats in captivity. Animals used to test food source were as follows: a, *Bathyergus suillus*; b, *Cryptomys h. hottentotus*; c, *Cryptomys damarensis*; d, *Cryptomys janetta*; e, *Bathyergus janetta* (after Bennett & Jarvis, 1995).

Food type	Energy value	Total food water content	Ash	Fibre content dry mass (% ± S.D.)		Faecal water content	n
	(kJ g^{-1})	(%)	(%)	Food	Faecal material	(% ± S.D.)	
Couch grass [a]	15.9	70.4	13.0 ± 1.6	23.0 ± 1.8	21.5 ± 4.3	15.3 ± 1.9	(10)
Sweet potato [b]	15.5	75.0	7.0 ± 0.6	4.1 ± 0.4	–	4.4 ± 2.4	(8)
Gemsbok cucumber [c]	14.2	80.3	6.3 ± 2.7	25.2 ± 8.0	41.5 ± 1.3	23.5 ± 7.6	(14)
Wild onion corm [d]	15.3	77.0	10.0 ± 2.9	4.7 ± 1.1	–	2.3 ± 1.0	(7)
Tuber (*Herrea*) [e]	15.5	78.7	5.0 ± 1.0	6.4 ± 0.5	15.4 ± 5.0	13.4 ± 4.8	(5)

burrows and found that the fibrous central pith of the cucumbers was rarely eaten.

The digestibility and assimilation of food is dependent upon many parameters, including the energetic value of the food and, most importantly, the fibre content. Energy stored in bulbs and corms is very concentrated and the fibre content is low, whereas some tubers and above-ground aerial vegetation have higher fibre content and are therefore less digestible (Table 3.2). Couch grass and the tuberous gemsbok cucumber, *Acanthosicyos naudinianus*, (fed upon by Damaraland mole-rats) have similar fibre contents, which are five times as great as those recorded for bulbs. Despite their similar fibre content, it is noteworthy that while the coefficient of digestion for the Cape dune mole-rat is high for couch grass, this was not true of the gemsbok cucumber fed to Damaraland mole-rats. In contrast, the tubers eaten by the Namaqua dune mole-rat, *Bathyergus janetta*, have fibre contents (6%) comparable to those of the fleshy bulbs and corms (4%). Of the two species of *Cryptomys* fed on low-fibre geophytes, and all species of mole-rat fed upon sweet potato, only 3.3 to 4% of the ingested energy was lost in the faeces. In contrast, the two species of mole-rats maintained upon high-fibre foods showed higher energy loss in their faeces, but that of the dune mole-rat (8.2%; Table 3.1) was markedly less than that of the Damaraland mole-rat (45%).

The loss in body mass of some of the mole-rat species fed upon high-fibre foods may in part be the result of a faster rate of the passage of food through the gut. All African mole-rats studied to date have a similar gut morphology and the caecal length expressed as a percentage of the total gut length is constant (Broll, 1981). The faeces of Damaraland mole-rats fed on a high-fibre diet retained 23.5% of the water that could be obtained from the food source, whereas the same species fed upon the corms of *Dipcadi* lost only 2.3% of the total water in the faeces (Table 3.2). Thus, there would appear to be no special adaptations of the gut morphology in any of the species for high-fibre diets. The loss in body mass experienced by mole-rats fed on high-fibre diets may well be due to increased faecal water loss. The high humidity of the burrow would alleviate this physiological loss of water and retain the positive water balance of the animal.

3.2 THE DISTRIBUTION OF GEOPHYTES

The size, distribution and nearest neighbour distance between the

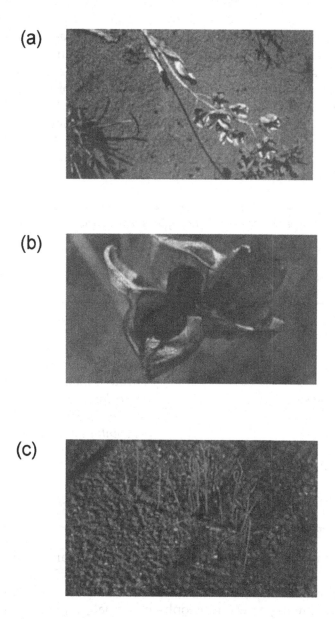

Figure 3.2. Mechanism of seed dispersal and subsequent clumping of geophytes in the habitat of *Cryptomys damarensis*, at Dordabis, Namibia, (a) general view of the seed pod of *Dipcadi glaucum*, showing layered, flattened black seeds; (b) magnified view; (c) clumped distribution of germinating seeds due to localised dispersion (photographs © Tim Jackson).

geophytes, together with the available digestible energy, may be an important correlate of sociality, an idea that we will discuss further at the end of this chapter and in Chapter 8. The distribution of geophytes is determined by both the rainfall characteristics of the habitat, and the nature of their mode of reproduction and seed dispersal (Figure 3.2). In mesic regions that are inhabited mainly by solitary mole-rats, there is high species diversity of geophytes (Du Toit *et al.*, 1985; Lovegrove and Jarvis, 1986; Table 3.3). The geophytes are generally small with a mean mass of less than 2 g, but have high coefficients of digestibility (up to 95% of the total energy stored in the bulb) aided by a relatively low fibre content (Du Toit *et al.*, 1985; Bennett and Jarvis, 1995).

Table 3.3. Geophytes consumed by *Georychus capensis* and *Cryptomys hottentotus hottentotus* inhabiting mesic regions (after Spinks, 1998; Lovegrove and Jarvis, 1986).

Family	Species
Hyacinthaceae	*Albuca cooperi*
	Lachenalia klinghartiana
	Ornithogalum secundum
	Ornithogalum thyrsoides
Iridaceae	*Hexaglottis virgata*
	Romulea rosea
	Micranthus junceus
	Homeria schleretchi
Oxalidaceae	*Oxalis* sp.
Aizoaceae	*Herrea blanda*
Asphodelaceae	*Trachyandra* sp.

In mesic regions, the mean nearest neighbour distance between two corms or bulbs is generally very close (between 0.38 and 0.87 m; Bennett, 1988). Consequently, the distance burrowed by a mole-rat before encountering a second geophyte is small, and the risk of not finding a geophyte is low. Similarly, the inter-clump distances are generally short (less than a metre) and allow solitary mole-rats to move from one clump to another without creating a deficit in the animal's energy budget. In addition to the rich diversity of geophytes in the mesic habitat, there is also a greater proportion of aerial vegetation such as grasses available for much of the year. The solitary species occupying this habitat have a more varied diet and undermine this aerial vegetation from the relative safety of their burrows (Davies and Jarvis, 1986; N.C. Bennett, pers. obs.). Indeed, studies investigating the gut contents of various solitary mole-rats have supported the visual observations of solitary mole-rats supplementing their diets with this food resource (Beviss-Challinor, 1980; Broll, 1981; Davies and Jarvis, 1986).

In the semi-arid and arid regions that are inhabited predominantly by social mole-rats, the species of geophytes are different from those found in mesic regions (Tables 3.4 and 3.5). The geophytes are also of a larger size than those encountered in mesic regions, but some, such as the gemsbok cucumber, have a lower coefficient of digestibility of around 50%, due to a higher fibre content (Bennett and Jarvis, 1995). Consequently this food resource provides less digestible energy gram for gram than do the geophytes in the mesic environment.

Table 3.4. Geophytes consumed by *C. damarensis* inhabiting an arid region (after Jarvis *et al.*, 1998, © Springer-Verlag).

Family	Species
Hyacinthaceae	*Dipcadi glaucum*
	Dipcadi platyphyllum
	Dipcadi mariothii
	Dipcadi bakeranum
	Ledebouria rerduta
	Ledebouria undulata
	Ornithogalum stapffii
Portulacaceae	*Talinum arnotii*
Eriospermaceae	*Eriospermum rautenenii*
Cucurbitaceae	*Acanthosicyos naudinianus*

The mean nearest neighbour distance between two gemsbok cucumbers in arid regions is also greater, with a mean value ranging from 1.9 to 2.4 m (Jarvis *et al.*, 1998), than geophytes in mesic regions. Thus, more effort is required by the mole-rat during the process of finding and harvesting the food. Calculation of the coefficient of dispersion, i.e. the variance in geophyte density divided by its mean, reveals that the gemsbok cucumber has a clumped or contagious distribution. Inter-clump distances of geophytes in the more arid regions may exceed 20m and are generally much greater than those encountered for geophytes in the mesic environment.

On depleting an area that is particularly rich in tubers, the mole-rats may have to burrow considerable distances to reach a new area with a similar density of tubers (Jarvis *et al.*, 1998). In the arid regions occupied by the naked mole-rat, the large tubers of *Pyrenacantha* sp., like *P. kaurabassana*, form the major food resource. These tubers average around 5 kg of food per tuber, but may sometimes reach a massive 30 kg, and therefore are a veritable feast when found (Brett, 1991a). However, *Pyrenacantha* tubers are randomly distributed and the inter-tuber distances can be relatively high, making them hard to find. In one plot examined by Brett (1991a), the mean nearest neighbour distance was 3.3 m. Likewise, another species eaten by naked mole-rats, *Macrotyloma maranguense*, may occur in large patches, but nearest neighbour distances of these 'food-rich' patches may average over 30 m. Therefore, to find this widely dispersed food necessitates cooperation in burrowing by the mole-rats (Brett, 1991a).

Table 3.5. Geophytes consumed by *Heterocephalus glaber* inhabiting an arid region (after Brett, 1991a).

Family	Species
Icacinaceae	*Pyrenacantha kaurabassana*
Leguminoseae	*Macrotyloma maranguense*
	Vigna friesiorum
	Vigna membranacea
Cucurbitaceae	*Dactyliandra stefaninii*
	Coccinia microphylla
Iridaceae	*Anthericum venulosum*
Portulacaceae	*Talinum caffrum*
Araceae	*Stylochiton salaamicus*
Cyperaceae	*Cyperus riveus*
Gramineae	*Chloris roxburgiana*
Acanthaceae	*Thunbergia geurkeana*
Convolvulaceae	*Ipomoea batatus*

The semi-arid and arid regions occupied by mole-rats also contain smaller bulbous geophytes, but again these are usually greater in size (>5 g) than those found in the mesic areas. These geophytes still retain similar dispersion patterns and coefficients of digestibility to the geophytes in the mesic areas. The gross energy values of the edible fractions of the various bulbs and tubers utilised by mole-rats inhabiting the arid regions varied between 14.2 and 15.3 kJ g^{-1} dry weight, and are similar to those in mesic areas where values are 14.3 to 17.6 kJ g^{-1} dry weight. Although the energy values per gram dry weight are slightly less in the geophytes from arid environments, the overall total energy content of these larger geophytes is greater than that of the small bulbs of the mesic regions.

The majority of studies examining resource characteristics and their role in the evolution of sociality in the Bathyergidae have concentrated on single sites, and regional variation has not been considered until recently. Jarvis *et al.* (1998) have investigated two distinct areas occupied by the Damaraland mole-rat. The different findings observed in these areas highlight some of the potential problems of making species-level generalisations from limited field studies. At Dordabis, Namibia, geophytes are small, numerous and with inter-clump distances less than in the Kalahari, South Africa, where geophytes are also larger and less numerous. The data from long-term mark–recapture studies in Dordabis (Bennett and Jarvis, unpubl.) and from a study by Lovegrove and Knight Eloff (1988) in the Kalahari, revealed a similar home range for large colonies of mole-rats: 10,000 m^2 in the former and 13,000 m^2 in the latter. Home ranges of this size imply that the distribution of small clumps is much less important than large-scale clumping. In the small colonies, or established pairs, of the Damaraland mole-rat, small-scale clumping probably plays a more important role than for large-sized colonies.

Within the same dune field of the Kalahari Gemsbok Park, there were marked differences in the resource characteristics in two independent studies. At Twee Rivieren, the gemsbok cucumbers occurred in clumps of around 25 m^2 spaced 20 to 25 m apart. In contrast, at Nossob, Namibia, 100 km due north of the site of Jarvis and Bennett at Dordabis, Lovegrove and Knight-Eloff (1988) noted that gemsbok cucumbers occurred randomly in very large patches (0.25 km^2) separated from each other by distances of more than 0.5 km. Inter-clump distances are not only an important factor for a colony as it forages, but also during times of dispersal and formation of new colonies from founding pairs. At the aforementioned study site of Lovegrove and Knight-Eloff (1988), while a colony would probably survive in the confines of a large patch, potential problems could arise during dispersal as the large inter-clump distances mean emigrating individuals could be left out in a region devoid of food. At Dordabis, where the geophytes are much smaller and the clumps very much closer together, the risk of starvation to the mole-rats is much lower than the more widely spaced gemsbok cucumbers. Consequently, in this environment, the probability of successful dispersal would be much greater as the chances of harvesting food are more favourable. Lone dispersal of animals would be possible whilst coalitions of dispersing animals should have a high probability of success.

3.3 FORAGING METHODS AND OPTIMALITY THEORY

Optimal foraging theory assumes that natural selection has shaped the way in which animals forage, such that the rate at which energy is acquired and consumed is maximised (Pyke, 1984; Pyke *et al.*, 1977). An important extension to foraging theory is that of 'optimal central place foraging', in which organisms collect food and carry it to the central store or nest in a way which maximises the rate at which energy is delivered (Orians and Pearson, 1979; Andersson, 1981). Barnett (1991), in a study of Damaraland mole-rats, found that the handling time for geophytes decreased proportionally with an increase in geophyte size. The handling time of the geophyte is energetically very expensive and is determined by both the size and nature of the food resource (Hughes, 1979; Kaufman and Collier, 1981). It follows that a mole-rat capable of anticipating whether to discard or to not completely eat a geophyte (an energy saving behaviour) could reduce handling costs by only partially removing the husk of the geophyte prior to discarding (Barnett, 1991). Thus, incomplete removal of the husk from the larger geophytes may be considered an adaptive energy strategy. Optimality theory suggests that foragers should ideally choose diets which maximise the net yield of energy per unit time (Charnov, 1976; Pyke *et al.*, 1977). From her regression analysis, Barnett (1991) showed that, for the Damaraland mole-rat, energy gained per unit time was greater when eating larger geophytes irrespective of the size of the animal and, for all sized animals, smaller geophytes were found to be the least efficient to consume. However, Barnett (1991) also found that the small geophytes were actually preferentially consumed both at the digging site (*in situ*) and in the food store. This presents us with an anomaly: why should the mole-rats have preferentially consumed the smallest and least profitable resource? It is probable that the selective pressure shaping the responses to geophytes is not one of energy requirements for consumption, but rather a function of storage optimality, the main tenet on which central place foraging is based (Orians and Pearson, 1979). Now, if the energy gained by storing an item of food is only slightly greater than that of performing the caching trip, then from a costs and benefits point of view, it would be more logical not to store the food item, but rather to consume it. Royama (1970) has shown in birds that the mean size of food items carried to the nestlings to feed them is greater than the size of food consumed by the provisioning bird. The selective storage of larger geophytes in preference to small ones by mole-rats is similar to this. The 'shelf life' and palatability of the different bulb types probably play an important role in the selection of bulbs, with medium and large-sized geophytes being preferentially stored because they have a longer shelf life.

To look at foraging in a controlled way, behavioural trials were conducted in captivity upon functionally complete colonies of Damaraland and common mole-rats. Animals were placed in a central arena from which radiated six tunnels leading to 'foraging trays'. These trays contained moistened soil into which either small, medium or large bulbs were placed, at differing densities, or blank trays which served as resource-deficient areas. The naive, satiated mole-rats were allowed into the central area from where they started their foraging activities. From these foraging trials it was found that mole-rats foraged indiscriminantly in terms of which trays they visited, and seemed unable to make decisions about the profitability of the resources contained within them. The test animals showed no differential response or any evidence of increased familiarity to a particular geophyte treatment as the experiment progressed (S.R. Telford, N.C. Bennett and J.U.M. Jarvis, unpubl.). According to the marginal value theorem (Charnov, 1976), high-quality geophyte patches should be exploited more readily than geophyte-poor areas. However, in the mole-rat foraging trials, the blank food trays containing no geophytes were still worked by the mole-rats, although they were exploited to a lesser degree than the trays containing geophytes. Of the trays containing geophytes of different densities and sizes, however, there was no obvious preference. It is thus plausible that whilst mole-rats are foraging, they do not make decisions concerning the profitability of differing resources. Indeed, the failure of mole-rats to dicriminate between geophytes of varying densities and sizes may well reflect the importance of establishing foraging runs and hence home ranges containing a resource. This would be particularly crucial in solitary species and social species in which pairs of mole-rats set up new colonies.

The establishment of a large home range that encompasses food resources in the wild is probably of paramount importance to the survival of the colony. If one considers that the time available to excavate extensive foraging burrows is limited by periods of rainfall which moisten the soil and herald mass excavation, it seems logical that vigorous excavation should take precedence over collecting the geophytes. It is noteworthy that in the laboratory foraging experiments, once a particular mole-rat had initiated digging in a particular tray, it usually remained faithful to this area. It could be argued that this is because prolonged excavation at one site is energetically more cost-effective than migration to and from different trays. On the other hand, there is no disadvantage in alternation of search effort between sites of high rewards and sites of unknown quality, as long as there is minimum risk of losing further rewards at the profitable region as well as a marginal possibility of encountering food within the unknown sites. The foraging burrows are such that mole-rats can leave areas and later return to these profitable areas without having to repeat the entire search process.

Recently, an interesting phenomenon has been recorded in colonies of naked mole-rats in captivity, which illustrates the importance of social learning in foraging behaviour (Judd and Sherman, 1996). In a way that bears an interesting similarity to eusocial bees dancing to communicate the whereabouts of food, it has been shown that naked mole-rats recruit colony mates to food sources, apparently by laying down an odour trail. In captivity, naked mole-rats are kept in artificial burrow systems mimicking those in the wild (albeit on a much smaller scale), consisting of a network of interconnecting perspex tunnels, with perspex nest and toilet chambers, and other chambers where food is introduced. Judd and Sherman (1996) withheld food in captive colonies of naked mole-rats for around 16 hours. They then found that individual foragers, when finding a particular food source for the first time after food deprivation, usually gave a characteristic 'chirp-like' vocalisation on returning to the nest chamber (on 74% of occasions). These specific chirp vocalisations not only alert colony members that a food source is present, but may also be individual-specific, thus identifying the forager. Subsequently, other colony members are able to find a food source discovered by a colony mate by following its pathway through the tunnels of the artificial burrow system, irrespective of whether the direction of tunnelling was the same or had been changed by modifying the layout of the tunnels in the captive colony (i.e. by swapping the perspex tunnels around). Also, these pathway preferences were lost if the perspex tunnels were cleaned, implying that recruits follow an odour trail laid down by the forager, probably by skin contact between the forager and the tunnel surface. The source of odour may come from urine smeared over the head and shoulders by the feet, a characteristic behaviour exhibited by mole-rats following urination in the communal toilet chamber (Lacey *et al.*, 1991). The chirp vocalisation and associated odour cue means that an individual who has just successfully foraged is recognised and only this particular trail is followed amongst the myriad of other odours within the colony.

We have already noted that in the habitat of the naked mole-rat, food is patchily distributed and widely dispersed (Brett, 1991b), and the condition of the soil makes burrowing for food time-consuming, risky and energetically expensive (Lovegrove, 1991). Socially learnt foraging behaviour could therefore certainly have an adaptive function in the wild, aiding navigation through complex burrows with the minimum expenditure of energy, increasing the efficiency of foraging and thereby benefitting the colony and ultimately the inclusive fitness accrued by non-breeding colony members. Similar foraging trails have also been described for terrestrial rodents e.g. Norway rats, *Rattus norvegicus* (Galef and Buckley, 1996; reviewed by Judd and Sherman, 1996). Whether or not other species of mole-rats also employ such odour trials is at present unknown.

3.4 FORAGING IN THE WILD

After good rainfall, the soil at burrow depth becomes moist and can be easily worked by the mole-rats, and this acts as a cue to initiate a crescendo of burrowing activity. All members of the colony are mobilised into a highly efficient and organised 'super-organism' for the prime purpose of excavation and food resource acquisition. The effort involved in such frenzies of activity is quite impressive. As discussed earlier, Jarvis *et al.* (1998) found that a Damaraland mole-rat colony of 16 animals excavated 2.6 tonnes of soil in less than two months. J.U.M. Jarvis and N.C. Bennett (pers. obs.) have noted that during the initial phases of burrow excavation, the mole-rats concentrate on enlarging their burrow system and little, if any, harvesting of food items takes place. As the soil dries, these exploratory burrows are revisited by the mole-rats who harvest the geophytes by digging minor lateral side branches from the major artery of the foraging tunnel.

Founding colonies (usually the reproductive pair and their first litter) are unable to excavate foraging tunnels which are long enough to sustain the mole-rats through dry times. The foraging burrows are inadequate to provide enough surrounding geophytes for the dry season when burrowing is economically not viable. An extensive field study conducted by Jarvis and Bennett (1993), which is still in progress at the time of writing, has provided an opportunity to study the population biology of the Damaraland mole-rat during a prolonged period of drought. This has enabled a critical examination of the survival of both newly founded as well as established colonies under these unusually harsh conditions. During the drought, small colonies were forced to extend their foraging burrows in these dry conditions, when they would normally not be extensively digging. In these colonies, molehills of fine dry soil could be found which were obviously energetically very costly to produce. The mole-rats in these colonies would also have been more vulnerable to predation and, furthermore, small group size also requires the reproductive animals to engage in a greater proportion of the total work. It is possible that this heightened risk of predation of the breeding pair may well explain why several of these young colonies failed to survive.

In the naked mole-rat, the supply of food to any particular colony depends principally on the rate at which *Pyrenacantha kaurabassana* tubers are found, as these make up over 90% of the palatable biomass. It has been shown by Brett (1991a) that naked mole-rats increase their rate of discovery of food plants by changing their burrowing behaviour in response to particular food plants. Jarvis and Sale (1971) noted that increased branching of burrows took place in patches of smaller tubers, *Macrotyloma maranguense*. In contrast, no increase in branching was found where burrows met the larger and more randomly distributed tubers of *P. kaurabassana*. This change in burrowing pattern from a straight, relatively unbranching burrow to a more winding and branching pattern in small patches of clumped but abundant geophytes provides an elegant example of area-restricted searching as an optimal foraging strategy (Krebs, 1978).

Although area-restricted searching was found by Brett (1991a) in the naked mole-rat, Jarvis (1985) found no relationship between the distribution of food plants and the burrow configuration nor any evidence that naked mole-rats followed roots as cues to the location of other plants. However, at the study site of Jarvis (1985) the naked mole-rats fed upon long fleshy roots, the study site being devoid of large tubers. Hence, area-restricted searching may only be advantageous where the food source is concentrated in patches.

It is well known that food resources play an integral role in determining the particular pattern of burrowing in subterranean animals (Andersen, 1982, 1987; Reichman *et al.*, 1982). Supporters of optimal foraging theory have proposed that the movements and subsequent burrow configurations of particular foraging animals are a consequence of natural selection favouring phenotypes which exhibit behavioural traits that in turn minimise the relative movement costs in comparison to the benefits (Pyke, 1984). Thus, one would expect that subterranean rodent moles should modify their exact burrow architecture in different habitats corresponding to the particular patterns of geophyte distribution. Spinks (1998), in his elegant studies of the common mole-rat at two localities that varied dramatically in their rainfall patterns, showed conclusively that the arid population had burrow systems which were notably more linear and longer than those occurring in the mesic habitat. He suggests that common mole-rats foraging in arid areas are obliged to increase the length of their burrow system relative to their conspecifics occurring in the more mesic distributions because of the lower geophyte density and the greater inter-clump distances.

Lovegrove and Knight-Eloff (1988) commented that a definitive study was required to determine whether or not mole-rats search and locate tubers and corms on a purely random basis. Jarvis *et al.* (1998) showed, from field observations, that Damaraland mole-rats are unable to detect the whereabouts of clumps of geophytes. Transects along newly thrown up rows of mounds and slightly weathered mounds (hence earlier excavations) showed that mole-rats frequently missed geophyte-rich regions by less than a metre (Table 3.6). A comparison of areas either inhabited or not inhabited by mole-rats showed no significant difference in geophyte density, supporting the notion that mole-rats forage blindly. Further evidence for this comes from work by Reichman and Jarvis (1989), who showed that in both the common mole-rat and the Cape mole-rat there was no evidence to suggest that either species foraged in parts of the field that contained more plant material. They also showed that neither the common mole-rat, the geophyte specialist, nor the Cape dune mole-rat, which consumes significant quantities of above-ground plant material, chose areas rich in those particular components of the diet.

Table 3.6. A comparison of the number of geophytes adjacent to mounds/burrows, compared with the number situated away from mounds. Results are presented as the mean ± standard deviation (n in parenthesis). No significant differences were found between the two sets of quadrats in 10 transects, suggesting that mole-rats do not selectively forage towards resource-rich areas (Student's t-test; Jarvis et al., 1998).

Plot	Number of geophytes		p value
	Near mounds	Away from mounds	
1	6.89 ± 1.29 (9)	5.33 ± 2.57 (18)	> 0.3
2	4.50 ± 1.29 (4)	9.20 ± 6.61 (35)	> 0.1
3	4.22 ± 5.38 (9)	2.88 ± 3.64 (33)	> 0.3
4	22.25 ± 15.52 (4)	23.20 ± 11.14 (10)	> 0.8
5	3.92 ± 2.43 (12)	2.41 ± 2.49 (27)	> 0.08
6	7.30 ± 3.09 (10)	5.68 ± 3.23 (22)	> 0.1
7	5.60 ± 4.62 (10)	5.56 ± 4.67 (18)	> 0.09
8	14.83 ± 4.79 (6)	12.60 ± 10.10 (10)	> 0.6
9	12.77 ± 6.42 (13)	12.53 ± 8.64 (13)	> 0.9
10	13.11 ± 5.71 (9)	14.85 ± 11.71 (13)	> 0.6
Mean	8.79 ± 7.22 (86)	7.77 ± 8.11 (205)	> 0.3

Brett (1991b) suggested that the naked mole-rat burrowed in a blind but unidirectional fashion at different levels in the soil. Foraging blindly would be a reasonable foraging behaviour for discovering large P. kaurabassana tubers which are randomly distributed. He also suggests it is highly unlikely that naked mole-rats can sense the presence of tubers over long distances through the soil, i.e. by chemical means. It is more probable that foraging naked mole-rats can only detect geophytes at very short distances (<10 cm), perhaps by smell. It is more probable that the rootlets or root hairs provide a physical cue to the mole-rat of a nearby geophyte. Jarvis et al. (1998) have shown that a mole-rat at their Dordabis study site would have to dig blindly, but in a directional way, for an average of 1.79 m to encounter the next geophyte or clump of geophytes. A 130 g mole-rat would therefore have to dig an average of 10.7 m in length to find the 34 g of bulbs needed to sustain its daily resting metabolic rate. However, if the animal were continuing to exploit a single clump of bulbs it would only have to forage about 2.9 m. The relative linearity of the excavation is an energy conserving strategy in that mole-rats do not apparently search the same area twice.

Table 3.7. Densities of geophytes in three areas occupied by large *Cryptomys damarensis* colonies (>18 individuals), five areas occupied by small colonies and three areas lacking mole-rats. The mean number of geophytes differed significantly between the three ($p < 0.00001$; Jarvis *et al.*, 1998, © Springer-Verlag).

Colony size	Total number of plots	Mean density of geophytes (no. m^{-2})	Standard deviation	Range
Large	62	60.68	33.53	7–127
Small	97	49.30	26.89	5–121
No mole-rats	60	22.28	22.28	0–78

Interestingly, studies of the Damaraland mole-rat (Jarvis *et al.*, 1998) showed that, within an area contained by their burrow system, larger, established colonies occur in regions of higher geophyte densities compared to smaller-sized colonies where geophyte densities are lower. In turn, areas containing no mole-rats were the poorest in geophyte density (Table 3.7.).

3.5 THE FOOD STORE

Geophytes occurring in the food stores of mole-rats normally show species selectivity. It is unknown why mole-rats should store one particular species more readily than another, but possible reasons are the shelf life of the commodity, its nutritional value, its abundance in the veld or its palatability. In a study investigating geophyte species-specific abundance in the open field and the percentage abundance in the food store, Lovegrove and Jarvis (1986) found that the Cape mole-rat and the common mole-rat showed a definite pattern in the preferential storing of geophytes in the food store. If one compares the composition of species occurring in the soil of the habitat with those which have been stored in the food cache, it becomes readily

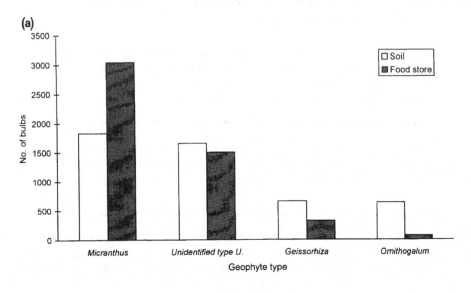

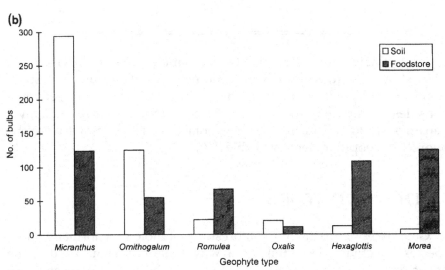

Figure 3.3. The number of geophytes excavated from the food store and the expected numbers sieved from quadrat samples from the burrow system of (a) the Cape mole-rat, *Georychus capensis*, and (b) the common mole-rat *Cryptomys h. hottentotus*. The numbers of geophytes in the food store are as counted, while those for the soil samples were obtained by multiplying the observed proportion of geophytes sieved from quadrats by the total number of geophytes found in the store (after Lovegrove and Jarvis, 1986).

apparent that there is selective storage of particular species of geophyte with respect to their relative abundance in the field (Figure 3.3).

The field observations of food stores support the laboratory foraging trials mentioned earlier, in that geophytes found within the food store had a significantly larger mean diameter compared with the mean diameter of bulbs collected in the immediate environs of the burrow system (Bennett, 1988; Figure 3.4). Again, this suggests that smaller geophytes were usually consumed on encounter, and that once the mole-rats are satiated these smaller geophytes were stored temporarily in the food store along with the larger-sized geophytes. Mole-rats typically excavate their foraging burrows at depths of less than 20 cm. Fewer smaller geophytes are likely to be encountered at this depth (Davies and Jarvis, 1986; Bennett, 1988; Jarvis *et al.*, 1998). These strategically positioned burrows may therefore represent an adaptation to the costs of searching, by enabling the mole-rats to thoroughly exploit foraging regions and minimise the encountering of smaller-sized geophytes which are energetically less profitable for harvesting.

The selective consumption of geophytes in the field naturally contributes to the differential composition of geophytes in the store. In the foraging trials on colonies of the Damaraland mole-rat described in Section 3.3, the selective consumption and storage of geophytes, which had been collected from foraging trays, was also investigated. After each trial, the numbers of each size category (small, medium and large) and the respective biomass were measured, and the geophytes placed back into the store. Many of the small geophytes were eaten *in situ*, but some were stored along with the other two size categories which were not preferentially eaten on encounter. Overnight the colony were allowed to visit the food store where they selectively ate the small and some of the medium-sized geophytes (S.R. Telford, N.C. Bennett and J.U.M. Jarvis, unpubl.).

There also appear to be differences in the absolute numbers of geophytes stored in caches in mesic and arid habitats (Spinks, 1998). In general, considerably fewer bulbs were stored by mole-rats at Steinkopf, South Africa (arid area). However, the significantly greater mean mass of the stored geophytes at Steinkopf meant that the total biomass of the stored geophytes here was much greater than at the mesic site. The diversity of stored geophytes was similar between the mesic and arid sites and, at both sites, the food caches were dominated by a single species of geophyte. Spinks (1998) showed, in the food stores of three separate colonies of the common mole-rat,

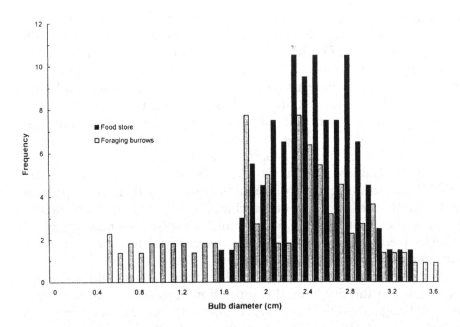

Figure 3.4. Size frequency distribution plots for geophytes (*Dipcadi* sp.) taken from a food store and from around 9 m of *Cryptomys damarensis* burrow (after Jarvis *et al.,* 1998, © Springer-Verlag).

that the size frequencies of the geophytes in the store differed significantly from random quadrats around the same burrow system (Table 3.8). This is very strong evidence of either a preference to store a particularly palatable geophyte or to store a geophyte which has a particularly long shelf life. In reality, it is probably the former which is true. Why do mole-rats store geophytes? Interestingly, food stores typically represent small energetic caches and unless they are continuously replenished they can only maintain the energy requirements of the colony for around five to 10 days. It is possible that the food store is utilised by the breeding female when she has young or as an emergency food reserve for the colony. The food store is usually positioned close to the nest area and this tends to lend support for its utilisation by the breeding female during lactation and early post-natal development of the pups, when her presence close to the young is crucial.

Table 3.8. In *C. h. hottentotus*, Spinks (1998) has shown at Steinkopf that food stores were composed almost entirely of *Ornithogalum secundum* (>90%), while a second store contained >60% of *Lachenalia klinghartiana*. In all these systems examined, the size frequencies of geophytes in the food stores differed significantly from random quadrats around the same burrow systems. The food stores contained fewer geophytes in the smaller size categories and more geophytes in the larger size category (from Spinks, 1998).

Food store	Size class (g)	Distribution of geophytes		χ^2	Significance
		Food store	Random		
1	0.00–0.25	0	83		
	0.25–0.50	0	66	558.2	$p < 0.00001$
	0.50–1.80	47	65		
	>1.80	236	69		
2	0.0–1.1	5	43		
	1.1–2.4	36	37	60.6	$p < 0.00001$
	2.4–4.7	41	38		
	>4.7	71	38		
3	0.0–0.2	1	43		
	0.2–0.3	1	35	484.9	$p < 0.00001$
	0.3–0.5	8	49		
	>0.5	155	37		

3.6 ECOLOGICAL CONSEQUENCES OF BURROWING BY MOLE-RATS

Many herbivores have a dramatic impact on plant survival, growth and reproduction. However, little is known about how subterranean vertebrate herbivores influence plants, directly or indirectly (Andersen, 1987). The labyrinthine burrows of mole-rats are normally in a state of constant flux, and the soil substrate is thus also constantly being modified and this can affect the surrounding plants. Likewise, 'browsing' damage to the underground storage organs of geophytes can dramatically affect the survival of such plants, may promote the spread of certain geophytes, and in some cases may promote growth as a result of the regeneration of cormlets and bulbs (N.C. Bennett and J.U.M. Jarvis, pers. obs.).

As well as consuming both above- and below-ground plants, mole-rats can dramatically alter the soil in which they occur by the construction of their extensive burrow systems and the subsequent deposition of large amounts of excavated soil onto the surface as a series of mounds. North American pocket gophers have been found to have a dramatic effect on the landscape. The burrows of the gophers may underlie 5 to 10% of a field (Andersen, 1982; Reichman *et al.*, 1982) whilst mounds may cover between 8 to 28% of the field surface (Grant *et al.*, 1980; Spencer *et al.*, 1985; Reichman and Jarvis, 1989). Such alterations to the soil can have a considerable impact on the plant biomass within the field. Indeed, it may affect the success of particular plant species, since some plants become smothered by the extruded mounds. There is limited evidence to suggest that the mining activity of rodents may provide sites for plant colonisers (Platt, 1975; Reichman, 1988; Reichman and Smith, 1985). Similarly, the consistent shunting of soil from the subsurface onto the surface can have dramatic affects on nutrient turnover and distribution (Inouye *et al.*, 1987a,b). The selective foraging and consumption of particular species of geophytes can also dramatically affect the species richness of particular fields in which the rodent moles occur. The activities of these burrowing rodents thus ensure that there is abundant soil mixing, promote decomposition of covered vegetation, aerate the soil, and seem to assist in normal nutrient cycling in the soil (Platt, 1975).

A case study conducted by Reichman and Jarvis (1989) investigated the effect that three sympatric herbivorous mole-rats (the Cape dune mole-rat, the Cape mole-rat and the common mole-rat) had upon both the above- and below-ground vegetation in the western Cape of South Africa. The three species vary dramatically in mean body size, the solitary Cape dune mole-rat being around 850 g, the solitary Cape mole-rat about 180 g, and the social common mole-rat about 70 g. The Cape dune mole-rat consumes by far the largest amount of above-ground vegetation (50% of its diet). In contrast, the Cape mole-rat and common mole-rat consume almost exclusively below-ground plant material (Broll, 1981). Reichman and Jarvis have calculated that three sympatrically occurring mole-rats in the southwestern Cape affected almost 40% of the study site through the actions of their burrow excavation and mound production. Mole-rats can drastically affect the local landscape and it has been suggested by Cox (1984) and Lovegrove and Siegfried (1986) that over long periods of time their activities may well be important in the formation of the large structures known as mima mounds or Heuweltjies. The landscape of sections of the western Cape Province of South Africa is characterised by the presence of thousands of these circular, raised earth mounds, which are typically about a metre at their central axis and have a diameter of at least 30 m. The mounds are common in the wheatlands from Picketberg on the western Coast through to Stellenbosch, a small town on the outskirts of Cape Town, but tend to be restricted in the inland mountain valleys. Lovegrove and Siegfried (1986) have proposed that the origin of these mounds may be the result of the nesting behaviour of the harvester termite *Microhodotermes viator*, in addition to the activities of the common mole-rat. *M. viator* normally builds a spherical subterranean nest approximately 0.6 m high with a width of 0.8 m and which is at least 0.5 m below the surface. Lovegrove and Siegfried (1986) suggest that mole-rats, confronted with flooding of their burrows, opted to colonise the termite mounds, which in turn provided a flood-free refuge. After the rains, in the spring and summer, the inter-mound soils build up around the termite mounds as a result of lateral soil movement caused by the renewed burrowing activity of mole-rats. This is substantiated by the findings of Cox (1984), who has shown that the burrowing activity of geomyid pocket gophers on mima mounds results in an overall translocation of soil towards the mound. Lovegrove and Siegfried (1986) have thus formulated the 'termite–mole-rat hypothesis' to explain the construction of the pimple-like structures which pepper specific shallow soils of southwestern Africa. This hypothesis states that 'mima mounds in southern Africa are the result of shallow soils, a long winter rainfall season and the burrowing activities of termites and mole-rats'. This is an interesting hypothesis, yet it still requires substantiating data to categorically dismiss the possibility of other geological phenomena resulting in this landscaping. One such scenario is the production of mima mounds as the result of localised retreating ice sheets which may produce regularly spaced piles of soil with particles of differing composition to the surrounding soil.

3.7 MOLE-RAT GEOPHYTE CO-ADAPTATION

Geophytes are potentially a valuable food resource in regions that have nutrient-poor soils. Cody *et al.* (1983) posed the question when referring to the geophytic plants occurring in the Mediterranean, 'do Mediterranean plants, especially on nutrient-poor soils, "safeguard" their supplies of nutrients by being unpalatable to consumers, both invertebrate and vertebrate?' Herbivory upon the parent storage organ of a geophyte, particularly during the dormant season, can result in a total loss of stored nutrients, together with the entire genetic stock of the plant. Hence, it is not unreasonable to expect that there has been strong selective pressure on geophytes to safeguard their survival both morphologically and biochemically. Lovegrove and Jarvis (1986) have found very strong evidence of co-adaptation between the food source and the mole-rat. The corm *Micranthus junceus* is of interest since the two sympatric mole-rats, the Cape mole-rat and the common mole-rat, both feed upon and store the geophyte. A disproportionately low number of geophytes occur in the food store of the common mole-rat compared with that of the food store of the Cape mole-rat, where large numbers are stored, but why should this be so? The most plausible scenario, put forward by Lovegrove and Jarvis (1986), is that the common mole-rat consumes a greater proportion of the corms *in situ* or along the burrow system than are stored in the cache.

Lovegrove and Jarvis (1986) went on to ask, 'what could possibly be the selective advantage of being highly palatable?' Like many African members of the Iridaceae, the food and nutrients storage unit is a corm. Its structure, however, differs in that the swollen stem portion of the corm does not constitute a single unit, but consists of six tunicated globose storage segments. At the first leaf node above the corm (but always undergound) are found clusters of up to eight small tunicated auxiliary cormlets which are weakly attached to the stem. The mole-rats feed on the corm, using their incisors to strip off the outer protective tunic. During this process some of the corm segments become dislodged and fall into the burrow. Once a single segment has been removed, the parent corm is dropped. The parent corm is then relocated to another section of the burrow. This process is then repeated with additional dislodgement of corm segments. Consequently, the corm invariably loses one or two segments and these segments are dispersed. Lovegrove and Jarvis (1986) have proposed that the high palatability of *M. junceus* induces mole-rats to eat many of the corms along the burrow system, hence maximising dispersion of segments during harvesting. The clusters of cormlets sited at the top of the parent corm serve as a back-up nutrient and genetic store in the event of the parent corm being consumed in its entirety.

This relationship between the mole-rat and the geophyte elegantly illustrates how two completely phylogenetically unrelated organisms can co-evolve to form a profitable relationship, the one organism providing a food source and the second organism ensuring widespread dispersal of the genetic material. Indeed, *Micranthus* has evolved a unique combination of gustatory and morphological adaptations in response to herbivory to promote the survival of the species. The corm safeguards its future by two forms of investment, firstly having out-of-reach clusters of cormlets to maintain the genetic bank, and secondly, being segmented and promoting herbivory so as to enable maximal dispersion by the consuming individual.

3.8 THE RISKS OF FORAGING FOR GEOPHYTES

Over the last decade, an increasing body of evidence has accumulated to suggest that the level of sociality exhibited in the Bathyergidae may be related to two ecological factors, namely habitat aridity and the subsequent type and distribution of their food resources (Jarvis, 1978; Bennett, 1988; Lovegrove and Wissel, 1988). This idea has become known as the food–aridity hypothesis (Jarvis *et al.*, 1994), and we will return to this in detail in Chapter 8. Clearly, the distribution of geophytes not only has a profound influence on the energy expended during foraging, but also on how foraging is undertaken (optimal foraging theory), and hence the risks of foraging unsuccessfully.

Lovegrove and Wissel (1988) constructed a mathematical model which quantified the risks of unproductive foraging as a function of the dispersion patterns of the geophytes and the foraging group size of the mole-rat colony. Jarvis and Sale (1971), Bennett (1988) and Lovegrove and Wissel (1988) have shown that as the habitat of the particular mole-rat species becomes more arid, so the species becomes progressively smaller in body mass, the size of the colonies larger and the degree of sociality more advanced. This trend in sociality is based in part on the distribution and the size of geophytes upon which the mole-rats feed (see Chapter 8).

In general, as the degree of aridity of the environment increases so the geophytes become larger and more widely dispersed (Figure 3.5). However, this is only a generalisation, as small geophytes can occur sympatrically with the larger tubers in very arid areas with both resources being readily consumed (Brett, 1986; Brett, 1991a,b; Jarvis *et al.*, 1998). Hence, mole-rats occurring in arid areas undertake much greater risks (in general) when burrowing, simply as a function of the much lower probability of successfully encountering a geophyte. Once the tuber is encountered, it becomes a common resource to the colony. Lovegrove and Wissel (1988) have suggested that mole-rats cannot simply increase the number of animals foraging, since then the total energy demand in arid areas with limited resources would become excessive. A better strategy would be to increase the number of foragers, but to produce smaller-sized individuals that require less energy.

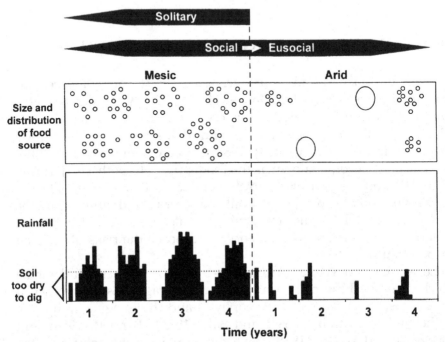

Figure 3.5. A schematic representation of the ecological constraints influencing group size in the Bathyergidae. In arid regions, there are limited windows of opportunity to burrow and extend the foraging tunnels. The dotted line represents the minimum rainfall required to promote burrowing (modified from Jarvis *et al.*, 1994).

Lovegrove and Wissel's model revealed that, by reducing body size and producing a larger number of smaller-sized foragers, the foraging risk was reduced in a geometric relationship. Lovegrove (1986b) made one of the most important findings in bathyergid energetics by showing that the arid-adapted Damaraland mole-rat had a mass-specific resting metabolic rate 43% less than the predicted value. Lovegrove also reported that resting metabolic rate in the Bathyergidae does not scale in the same way with body mass as other subterranean families, but is mass independent. This property of bathyergid mole-rats could therefore promote an increase in the number of foragers with a small body size in arid regions. Lovegrove and Wissel (1988) termed this property 'risk-sensitive metabolism'.

Table 3.9. A comparison of food resource parameters for selected species of the Bathyergidae (M = mesic habitat, A = arid habitat; modified from Spinks, 1998).

Species	Social status	Locality	Density of geophytes (no. m⁻²)	Biomass (g m⁻²)	Tot. energy available (kJ m⁻²)	Distrib. of geophytes	Source
Georychus capensis	Solitary[M]	Western Cape	100–500	120–600	480–2000	Clumped	Du Toit et al., 1985; Bennett, 1988
Cryptomys h. hottentotus	Social[M]	Western Cape	56	284.8	1068	Clumped	Spinks, 1998
Cryptomys h. hottentotus	Social[A]	Northern Cape	76	329.2	1434	Clumped	Spinks, 1998
Cryptomys damarensis	Eusocial[A]	Nossob, Kalahari	40–118	54.4–118	185–416	–	Lovegrove and Knight-Eloff, 1988
Cryptomys damarensis	Eusocial[A]	Dune, Kalahari	0.165	110.5	310.0	Random	Lovegrove, 1987
Cryptomys damarensis	Eusocial[A]	Twee Rivieren, Kalahari	0.03–0.63	12.7–257.4	35.6–721	Clumped	Jarvis et al., 1998; Bennett and Jarvis, 1995
Cryptomys damarensis	Eusocial[A]	Dordabis, Namibia	4–160	98	345.5	Clumped	Jarvis et al., 1998
Heterocephalus glaber	Eusocial[A]	S. Kenya	0.059	311.5	1401	Random	Brett, 1991a

The units in the header are in no. m^{-2}, g m^{-2}, and kJ m^{-2}.

The biomass of palatable geophytes available to different species of mole-rats (solitary through to eusocial) does not vary with aridity (Table 3.9). Bennett (1988) and Jarvis *et al.* (1994) have suggested that it is the pattern and density of resource dispersion, and not the total available biomass (and hence energy), which have driven the evolution of sociality in the Bathyergidae. This is evident from a number of surveys in the wild (Table 3.9; Chapter 8) where the habitats of different species have been compared, and arid regions shown to have lower densities of geophytes. Furthermore, an intra-species study by Spinks (1998) on the common mole-rat in the most mesic and arid regions of its range revealed considerably more geophytes at the mesic study site (Sir Lowry's Pass, Western Cape, South Africa) when compared to the arid site (Steinkopf, Northern Cape, South Africa).

According to the model of Lovegrove and Wissel (1988) and Lovegrove (1991), coloniality and group foraging reduce the risk of poor foraging performance whilst excavating for widely dispersed geophytes. As a result of the lower resource densities at the Steinkopf site, considerable pressure for group living and cooperative foraging will be placed on the species occurring in the more arid distributional ranges. Even so, the geophytes at both study sites exhibited a significantly clumped pattern of dispersion (Table 3.9). This is a consequence of the mode of reproduction and subsequent dispersal patterns exhibited by these plants (Figure 3.2). Indeed, food density is low at almost all localities frequented by mole-rats (be it mesic or arid), but the degree of clumping varies. The lower geophyte densities at Steinkopf, and indeed arid areas in general, may exaggerate clumping, resulting in greater mean inter-patch distances and subsequent lower resource densities within a given patch (Spinks, 1998).

The size of individual palatable geophytes may also be an important factor shaping sociality of the bathyergid mole-rats. Jarvis and Bennett (1991) note that the benefits accrued from cooperative foraging in arid environments will only pay off if sufficiently sized food items are secured to satisfy the energy requirements of the whole colony. The size of the food item is therefore critical, and a decrease in geophyte density in arid areas must be offset by an increase in the average size of the food items. Thus, the larger geophytes observed at Steinkopf enable the cooperatively foraging colonies to meet their energy requirements (Spinks, 1998).

The naked mole-rat at Mtito Andei, Kenya (Brett, 1991a) and Damaraland mole-rat at Twee Rivieren and Nossob, Kalahari

National Park, South Africa (Jarvis *et al.*, 1998; Lovegrove and Painting, 1987) have been shown to 'farm' the large geophytes that occur in these areas. In the case of the larger bulbs consumed by the common mole-rat at Steinkopf, although the largest geophytes here were markedly smaller than the tubers of *Pyrenacantha* or *Acanthosicyos*, the mole-rats nevertheless farmed them in a similar fashion at this site. Again, these farmed geophytes were too large to be transported effectively to the food store (Spinks, 1998).

In conclusion, it would appear that the African mole-rats have taken full advantage of the subterranean niche and its associated food reserves, in terms of their physiological and behavioural adaptations. However, as we have seen, there are costs and risks involved in this lifestyle. Perhaps the most unique feature of many of the species of mole-rats is their complex social behaviour, a major component of which is cooperation in foraging and the exploitation of the food resource. We will now consider in detail the nature and spectrum of social organisation in the family, and how this may have evolved in response to environmental constraints such as aridity, food distribution, burrowing costs and foraging risks.

Chapter 4

Social organisation in African mole-rats

4.1 SOLITARY OR SOCIAL?

The subterranean niche imposes very similar constraints on the different species that inhabit it (Nevo, 1979) and, as a consequence, there is considerable parallel evolution, or convergence, amongst subterranean animals with divergent evolutionary histories. The Bathyergidae is of particular interest since, although it shows much convergence with other subterranean families, it also shows a number of notable exceptions, amongst which is the degree of sociality exhibited by the family (Chapter 1; Jarvis and Bennett, 1990).

Of the seven families and sub-families of rodents containing some 26 genera of subterranean rodents (Table 4.1), apparently only the Bathyergidae has members that exhibit a spectrum of social organisation ranging from strictly solitary to cooperative breeding and eusociality. A characteristic feature of the social bathyergids is the restriction of reproduction to a single breeding female and a small number of males, with the remaining colony members being reproductively quiescent (Jarvis, 1981, 1991; Bennett and Jarvis, 1988a; Bennett, 1989; Bennett et al., 1994a,c; Bennett and Aguilar, 1995). This characteristic socially-induced reproductive suppression is discussed fully in Chapter 6.

Although a few species of subterranean rodents are still to be fully studied in the wild, it is perhaps surprising that in the other six families, convergent evolution does not seem to have led to the evolution of cooperative breeding systems like those seen in the Bathyergidae. This is presumably due to differences in environmental constraints and selective pressures in the habitats of these other families.

Table 4.1. A summary of the families of subterranean rodents and the genera that they contain. * denotes genera not modified for fossorial life but may burrow (modified from Jarvis and Bennett, 1991).

Family	Genera	Distribution
Bathyergidae (African mole-rats)	*Heterocephalus Heliophobius Georychus Bathyergus Cryptomys*	Sub-Saharan Africa
Rhizomyidae (bamboo/root rats)	*Tachyoryctes Rhyzomys Cannomys*	Africa, Asia
Spalacidae (mole-rats)	*Spalax*	Europe, Asia, North Africa
Octodontidae (octodonts/degus)	*Octodon* Octodontomys* Octomys* Aconaemys Spalacopus*	South America
Ctenomyidae (tuco-tucos)	*Ctenomys*	South America
Geomyidae (pocket gophers)	*Thomomys Geomys Heterogeomys Macrogeomys Orthogeomys Pappogeomys Zygogeomys Cratogeomys*	North America
Cricetidae (voles)	*Myospalax Ellobius Prometheomys*	Asia, Eurasia

Amongst the other subterranean rodents, several species of the family Ctenomyidae (tuco-tucos), including *Ctenomys peruanus* (Pearson, 1959) and *C. sociabilis* (Pearson and Christie, 1985), exhibit multiple occupancy of burrows. However, neither the duration of this cohabitation nor the organisation within the group has been documented. It appears that unlike the social Bathyergidae, *C. peruanus* may have a group of several sexually active females sharing a burrow, whereas the male is highly territorial and retains a strictly solitary existence (Pearson, 1959; Pearson and Christie, 1985). The curoro, *Spalacopus cyanus*, occurs in the semi-arid region of central Chile. It reportedly occurs in colonies of up to 15 individuals (Reig, 1970) and, again, there is some evidence of more than one sexually active female in the colony. However, the degree of social organisation is poorly understood and further work is required to investigate the possibility of *Spalacopus* being merely gregarious.

Multiple occupancy of burrows in solitary species occurs only briefly, during the breeding season or when the female has young (Bennett & Jarvis, 1988b; Altuna *et al.*, 1991). In marked contrast, the social species remain as a discrete colony unit for much of their lives. The original founding pair forms the nascent colony, which then increases in size through the addition of their offspring. The colony members cooperate in foraging and performing burrow maintenance activities that benefit all members of the colony and, in this respect, there are many parallels with the social insect societies (Chapter 8).

The social Bathyergidae characteristically share a burrow system and live together throughout the year. The colonies are familial units which show a marked reproductive division of labour, in that a single reproductive female is solely responsible for the production of offspring. The reproductive female retains her position for her whole lifetime and the remaining colony members are reproductively quiescent whilst resident in her colony (Jarvis and Bennett, 1993; Jarvis *et al.*, 1994). The bathyergids are unusual amongst vertebrate cooperative breeders in that at least two species, the Damaraland mole-rat, *Cryptomys damarensis*, and the naked mole-rat, *Heterocephalus glaber*, fulfil the traditional three criteria for eusociality, i.e. reproductive division of labour, overlap of generations and cooperative care of the young (Jarvis, 1981; Jarvis and Bennett, 1993). Eusociality was previously thought to be restricted to social insect societies, such as termites, bees and ants (Batra, 1966).

Species of *Cryptomys* occur in mesic, semi-arid and arid habitats in the tropics as well as in more temperate regions. The degree of sociality ranges from pairs in the Natal mole-rat, *C. hottentotus natalensis* (Hickman, 1979; 1982), and small colonies, such as in the common mole-rat (Bennett, 1989) and the Mashona mole-rat, *C. darlingi* (Bennett *et al.*, 1994a), to the Damaraland mole-rat where colonies may occasionally number 41 animals (Bennett and Jarvis, 1988b; Bennett, 1990; Jacobs *et al.*, 1991; Jarvis and Bennett, 1993).

In this chapter we will concentrate our attention on the three species of social mole-rat which have been examined in the greatest detail in long-term field and laboratory studies, i.e. the naked mole-rat, the Damaraland mole-rat and the common mole-rat, *Cryptomys h. hottentotus*. Field studies conducted on these species have shown the longevities of the breeding pair of a colony can be more than four years in the common mole-rat (Spinks, 1998), exceeding eight years in the Damaraland mole-rat (J.U.M. Jarvis and N.C. Bennett, unpubl.), and at least 10 years in the naked mole-rat (S.H. Braude, pers. comm.).

4.2 COLONY SIZE

In social mole-rats, colony size depends on a number of factors, some of which reflect the degree of risk that the colony members face if they attempt to disperse and establish their own colony. Thus, in a long-term population study on the Damaraland mole-rat, colonies increased in size during long dry spells when the risks to dispersal were high. Then, following rainfall, numbers would drop as individuals dispersed (N.C. Bennett and J.U.M. Jarvis, unpubl. data). This was also the time when small, newly founded colonies appeared. These dynamic fluctuations in individual colony size and seasonal increases in small nascent colonies make it difficult to give precise figures for colony sizes for a species. Other factors that will affect colony size, and the rate at which a colony can increase in size, will be the amount of predation and life history parameters such as the number of pups in a litter and the length of gestation and inter-birth interval.

Naked mole-rats have the largest colonies of all the Bathyergidae (Figure 4.1), with a mean of around 80 animals (Brett, 1991b) although a few colonies of almost 300 have been captured by Brett (1991b) and Braude (1991; pers. comm.). Naked mole-rat litters are also the largest of the bathyergids with a mean of 12 and a maximum of 27 (Brett, 1991b; Jarvis, 1991).

Colonial mole-rats in the genus *Cryptomys* show a range of colony sizes but rarely exceed 14 animals. In all species, litter sizes are much

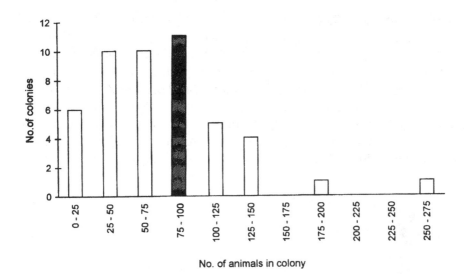

Figure 4.1. The size distribution of 48 complete living colonies of the naked mole-rat, *Heterocephalus glaber*, captured in the wild. The shaded histogram represents the modal colony size (adapted from Braude, 1991).

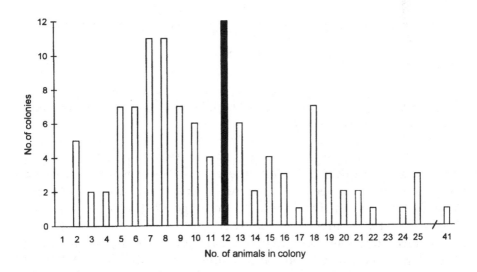

Figure 4.2. The size of 110 complete colonies of the Damaraland mole-rat, *Cryptomys damarensis*, captured in the wild. The shaded bar represents the modal colony size (data from J.U.M.Jarvis and N.C. Bennett, unpubl.).

smaller than for naked mole-rats, with means ranging from two to four, with a maximum of six pups in a litter (Bennett, 1989; Bennett *et al.*, 1994a; Gabathuler *et al.*, 1996; Wallace and Bennett, 1998; Moolman *et al.*, 1998). Exceptions to the smaller colony sizes in the cryptomids occur in the eusocial Damaraland mole-rat, where there are two records of the capture colonies containing 41 animals (Jarvis and Bennett, 1993). However, colonies of this size are not common and, overall, this species has a mean colony size of 11, with a mode of 12 animals (Figure 4.2).

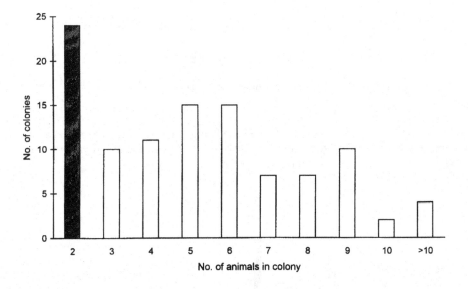

Figure 4.3. The size of 105 complete colonies of the common mole-rat, *Cryptomys hottentotus hottentotus*, captured in the wild. The shaded bar represents the modal colony size (data from Spinks, 1998)

Spinks (1998) examined colony size in two populations of the common mole-rat that inhabited different regions of South Africa. He found that, in this species, the degree of aridity of their habitat did not affect the mean (five animals) or range (2–14) of colony size (Figure 4.3). This similarity in colony size in the two study areas was unexpected as the risks of successfully dispersing would be expected to be higher in the arid habitat (mean annual rainfall 145 mm) than in the mesic part of their range (mean annual rainfall 652 mm). The data collection extended over three years and it is possible that the picture would have been clarified by a longer study, especially since the rainfall in the arid region was higher than average during the study period.

To date, there is less information on colony size in other species of *Cryptomys*. For five colonies of wild-captured Mashona mole-rats, the mean colony size was seven, ranging from five to nine animals (Bennett *et al.*, 1994a), while Hickman (1980) reported that colonies in Natal, South Africa, consisted of just a reproductive pair. H. Burda (pers. comm.) and Burda and Kawalika (1993) cite uncorroborated reports from local trappers that colonies of *Cryptomys* in Lusaka (probably *C. amatus*) may reach sizes of 26 animals, and those of the giant Zambian mole-rat (*C. mechowi*) may number as many as 60 (see page 18).

4.3 SIZE DISTRIBUTION AND COLONY BIOMASS

In the Bathyergidae, body size is influenced by a number of factors. Because the animals are burrowing through a dense medium, the energetic costs of extending the tunnels are very high and greatly exceed that of searching for food above ground. Vleck (1979, 1981) and Lovegrove (1989) have shown that the costs are linked to the burrow diameter, the soil type and the hardness of the soil. For this reason mole-rats do most of their burrowing when the soil is moist and easily worked and it also restricts large-sized mole-rats to the more easily worked softer soil types (Chapters 2 and 3). For example, the largest of the Bathyergidae, the solitary Cape dune mole-rat *Bathyergus suillus* occurs, as its name implies, in soft sandy soil, while the small naked mole-rat (mean mass 35 g) is found in areas with extremely hard soils.

The amount of energy available in an area can also affect body size and regional differences have been found in the mean body masses of both the naked mole-rat (Jarvis, 1985) and the common mole-rat (Spinks, 1998). Naked mole-rats occurring in the north of Kenya, in an area where food is not abundant, are smaller than those occurring in areas with more food. This food-related difference in body size even occurs within the same locality (Jarvis, 1985). Likewise, common mole-rats from a population in an arid part of their range are smaller than those from a region with higher rainfall and more abundant food. In both these examples, the smaller body size would reduce the overall energy expenditure of the colony without there being a decrease in the size of the colony workforce.

Sexual dimorphism is found in some, but not all, species of mole-rats. In the solitary dune mole-rats (the Cape dune mole-rat and the

Namaqua dune mole-rat, *B. janetta*) adult males are larger than the females and there is some indication that they may fight for mating rights. Indeed the males have extremely thick skin, under their chins and down the neck, which may provide some protection during fights for a mate. By contrast, for yet unknown reasons, there is no sexual dimorphism in the solitary Cape mole-rat (*Georychus capensis*).

In the social species the evidence for sexual dimorphism is somewhat obscured by the fact that status within the colony affects body size. Body size is not fixed on attaining adulthood and can change during the life of an adult animal. Thus, in naked mole-rats, the reproductive female is one of the largest individuals in the colony. On attaining her reproductive position she not only increases in mass but her vertebrae lengthen to give her a distinctively elongated body shape (Jarvis, 1991; O'Riain 1996). This probably enables her to carry a large number of foetuses without getting stuck in the narrow confines of the burrow. By contrast, when a male naked mole-rat becomes reproductively active, he may gradually lose mass and can often be identified as a thin, emaciated-looking animal (Jarvis, 1991). Amongst captive non-reproductive naked mole-rats, the first litters born to a colony typically attain the greatest body masses whereas animals from later litters are significantly smaller (O'Riain, 1996; O'Riain and Jarvis, 1998). Individuals that are colony defenders are large animals with robust jaw muscles and broad, strong incisors (O'Riain, 1996).

In *Cryptomys* the reproductive animals are amongst the largest colony members (Tables 4.2 to 4.4), but here the reproductive male is larger than the reproductive female and the most dominant animal in the colony. In the cryptomids there is again some evidence of size differences within the non-breeding members of the colony but this is not as marked as in the naked mole-rats. The common mole-rat and Damaraland mole-rat appear to exhibit sexual dimorphism based on body mass (Tables 4.2 and 4.4), but this is absent in the Mashona mole-rat (Table 4.3).

The sum of the body masses of a colony, i.e. its biomass, is a useful parameter to measure as it provides some indication of the total energy requirements of the colony (Tables 4.2 to 4.5). In many social mole-rats the biomass of an entire colony is considerably lower than that of a single dune mole-rat, in which adult males can occasionally exceed 2000 g (J.U.M. Jarvis, pers. comm.). This means, in effect, that the entire colony of social mole-rats can have similar overall energy requirements to a single solitary mole-rat and have the additional advantage that there are a number of individuals all

Table 4.2. Composition of three *Cryptomys hottentotus hottentotus* colonies in South Africa (after Bennett, 1989).

Location and colony	No. of mole-rats	No. of males	Mean mass of males ± S.D. (g)	No. of females	Mean mass of females ± S.D. (g)	Sex ratio M/F	Mean body mass ± S.D. (g)	Biomass (g)	Body weight of breeding female (g)
Darling 1	11	5	79.4 ± 33	6	41.1 ± 14	0.83	58.5 ± 30.5	643.5	–
Stellenbosch 1	13	8	56.9 ± 18	5	54.3 ± 16	1.60	55.9 ± 16.6	726.7	68
Stellenbosch 2	12	4	61.3 ± 22	8	45.3 ± 13	0.50	50.5 ± 18.0	607.0	71

Table 4.3. Composition of five *Cryptomys darlingi* colonies in Zimbabwe (after Bennett *et al.*, 1994a).

Location and colony	No. mole-rats	No. males >40 g	No. males <40 g	Mean body mass of males >40 g ± S.D. (g)	No. of females >40 g	No. of females <40 g	Mean body mass of females >40 g ± S.D. (g)	Mean body mass (g)	Total biomass (g)
Harare 1	9	4	0	72.0 ± 10.9	5	0	74.8 ± 13.7	73.5	662
Goromonzi 1	7	3	2	64.6 ± 17.2	2	0	57.5 ± 10.6	51.0	357
Goromonzi 2	7	4	2	73.2 ± 11.2	1	0	80.0 ± 0.0	62.1	435
Goromonzi 3	9	4	0	55.2 ± 16.9	5	0	51.9 ± 9.1	53.4	481
Goromonzi 4	5	3	0	60.0 ± 12.3	2	0	58.0 ± 14.6	59.2	296

Table 4.4. Composition of six *Cryptomys damarensis* colonies in Namibia (after Jacobs *et al.*, 1991).

Location and colony	No. mole-rats	No. males	Mean mass of males ± S.D. (g)	No. females	Mean mass of females ± S.D. (g)	Sex ratio M/F	Mean body mass ± S.D. (g)	Biomass (g)	Body mass of breeding female (g)
Otjiwarongo 1	22	15	103.8 ± 31.5	7	88.4 ± 14.6	2.14	98.9 ± 27.9	2089	–
Dordabis 1	12	5	130.0 ± 6.1	7	101.1 ± 31.4	0.71	113.1 ± 39.0	1358	152
Dordabis 2	25	11	185.5 ± 31.9	14	132.2 ± 32.6	0.78	155.7 ± 41.5	3893	206
Dordabis 3	17	6	117.3 ± 80.3	11	102.4 ± 40.6	0.54	107.7 ± 55.0	1831	176
Dordabis 4	16	7	202.1 ± 62.4	9	145.5 ± 39.7	0.77	170.3 ± 57.0	2724	205
Dordabis 5	15	8	146.3 ± 82.2	7	129.2 ± 54.5	1.14	138.3 ± 68.6	2075	220

Table 4.5. Composition of three colonies of *Heterocephalus glaber* from northern Kenya (Lerata) and four colonies from southern Kenya (Kathekani) (after Brett, 1991b).

Location and colony	No. mole-rats	No. males	No. females	Mean mass ± S.D. (g)	Sex ratio M/F	Biomass (g)	Body mass of breeding females (g)
Lerata 1	60	35	25	20.9 ± 5.2	1.40	1254	~
Lerata 2	40	21	19	17.7 ± 4.3	1.11	624	28
Lerata 4	82	45	37	27.9 ± 8.9	1.22	2111	52
Kathekani 2	52	26	26	39.1 ± 10.1	1.00	2034	61
Kathekani 7	93	46	47	30.7 ± 7.8	0.98	2857	54
Kathekani 8	70	38	32	33.1 ± 9.4	1.19	2315	59
Kathekani 10	25	10	15	20.4 ± 9.8	0.67	511	41

searching for the food that they need. This is an important consideration if the food is clumped and difficult to find (see Chapter 3). Thus, the mean colony biomass of the smaller-sized colonies of the social *Cryptomys* rarely attains a kilogram. In the common mole-rat the average colony biomass is around 660 ± 61 g (range 607–727 g) (Table 4.2; Bennett, 1989). Likewise in the Mashona mole-rat, colony biomass is small (446 ± 140 g; range 296–662 g) (Table 4.3; Bennett *et al.*, 1994a). The numerically larger colonies of the naked and Damaraland mole-rats typically have biomasses of about 2000 g (Tables 4.4 and 4.5).

4.4 CAPTURE ORDER WITHIN COLONIES

When capturing mole-rats in the wild, the burrows are opened up at active molehills, and animals are caught either as they come to re-seal their tunnels, or by means of a baited shutter trap placed at the opened section. An open burrow is a source of danger to mole-rats as it can potentially permit entry to predators such as snakes. It is therefore possible that an examination of the order in which colony members are captured can give us useful information on which animals respond to danger and which ones avoid these high-risk areas.

In a number of bathyergid species, the reproductive animals are often amongst the last animals to be captured. This suggests that they are least likely to visit areas of potential danger in the burrow (Brett, 1991b; Bennett, 1989; Jacobs *et al.*, 1991; Bennett *et al.*, 1994a). Beyond this, however, the evidence is ambiguous as to whether or not large-sized defenders are the first to come to a disturbed area. Thus, for example, significant negative correlations between body mass and capture order were found in five of the six colonies of naked mole-rats that were captured by Brett (1991b) but in only four out of 12 colonies captured by other workers (P.W. Sherman, unpubl.; Jarvis, 1985). Laboratory studies show that the larger naked mole-rats (with the exception of the breeders) in a colony are at the front-line in colony defence, and it is possible that the capture methods employed in the field do not elicit a defence response in the colony members (see Section 4.8).

In the genus *Cryptomys* no significant correlations were found between body mass and capture order in three colonies of the common mole-rat (Bennett, 1989) or in five colonies of Mashona and Damaraland mole-rats (Jacobs *et al.*, 1991; Bennett *et al.*, 1994a).

4.5 REPRODUCTIVE DIVISION OF LABOUR

Solomon and Getz (1997) report that there are essentially three basic hypotheses to explain reproductive suppression. Firstly, to decrease the number of reproductive females at a specific site; secondly, to prevent inbreeding; and thirdly, to delay pregnancy until female offspring are older and more likely to be successful in rearing the young. In the first hypothesis, the mother would benefit by exerting a restraining or suppressive effect on the reproduction of other females sharing the nest. If resources were limited, having numerous reproductive daughters might adversely affect the mother. The second hypothesis suggests that reproductive suppression of daughters occurring at the natal nest is adaptive for the daughter and her male relatives, if the only potential mates in the colony are the daughter's father and brothers, as is found in the Damaraland and the common mole-rat. Inbreeding avoidance may help to explain the situation arising in the species of *Cryptomys* studied to date, where inbreeding is avoided, but it cannot explain the reproductive division in the highly inbred naked mole-rat. In this species, reduced dispersal opportunities lead to continuous cycles of inbreeding, and outbreeding is thought to be uncommon (Faulkes *et al.*, 1990b, 1997a; Reeve *et al.*, 1990; Brett, 1991a; O'Riain *et al.*, 1996). The third hypothesis suggests that it may be adaptive for females to refrain from reproduction until after they have dispersed and established their own territory. Females who delay breeding should theoretically show an enhanced lifetime reproductive success when compared to females that do not delay their attempt to disperse. In the case of mole-rats inhabiting regions where ecological constraints on dispersal are high, it is easy to see how this could be important (see Chapter 8). We will return in detail to consider the proximate mechanisms involved in reproductive suppression in Chapter 6, and also consider how theories of optimal reproductive skew may further our understanding of the reproductive division of labour in social mole-rats.

As we have said in earlier chapters, all social mole-rats studied to date are characterised by the occurrence of a single reproductive female per colony (Jarvis, 1981, 1991; Bennett and Jarvis, 1988b; Bennett, 1989, 1990; Bennett *et al.*, 1994b,c; Bennett and Aguilar, 1995; Moolman *et al.*, 1998; Chapters 5 and 6). Although reproduction in the naked mole-rat is usually restricted to a single breeding female, in both the laboratory (Jarvis, 1991) and the field (Braude, 1991), dual-queen colonies are occasionally found and

Braude found evidence of the two females sharing the suckling of the pups. In the laboratory, however, the births of litters are rarely synchronised and the presence of two females in the colony leads to heightened bouts of aggression in the form of shoving and poor survival of the pups.

All colonies of wild-caught (>50 colonies) and captive (n = 65 colonies) Damaraland mole-rats have contained a single reproductive female (N.C. Bennett and J.U.M. Jarvis, unpubl.). However, Spinks (1998) has shown that of 42 wild-captured colonies of common mole-rats, two had two queens and the remaining (95%) were singularly breeding with one queen. The colonies containing two breeding females were only found in the more mesic part of their range.

The number of males involved in reproduction is more variable, and can be as many as three in the naked mole-rat (Jarvis, 1991; Lacey and Sherman, 1991), whereas in the genus *Cryptomys* it is usually a single male (Bennett and Jarvis, 1988b; Bennett, 1989; Bennett *et al.*, 1994a; Bennett and Aguilar, 1995). However, it is difficult to exclude the possibility of other 'non-breeding' males reproducing. In order to unambiguously confirm paternity within a colony, molecular genetic techniques such as DNA fingerprinting could be used, although to date this has only been done for the naked mole-rat. Faulkes *et al.* (1997b) have recently demonstrated that there can be multiple paternity by breeding males within a single litter of naked mole-rats in a captive colony, although the incidence of subordinate male reproduction remains unknown (see Chapter 5).

Mark–recapture studies on the naked mole-rat in Meru National Park, Kenya, suggest that replacement reproductives are drawn from within the colony. Braude (1991) found that three breeder replacements in wild-captured colonies were from females that had previously been non-reproductive members of the same colony. This contrasts with the Damaraland mole-rat where, in large established colonies, the death of a reproductive results in the colony breaking up the next time rainfall occurs. The colony members then disperse and attempt to locate an unrelated partner and found their own colony. Hence, the lifespan of a colony appears to be dependent upon the continued presence of the original reproductive pair of mole-rats (Jarvis and Bennett, 1993). Exceptions to this are in young, newly founded colonies where a foreign animal may join the small colony when one of the reproductives dies. The majority of newly founded colonies at the study site of Jarvis and Bennett (1993) in Namibia contained a pair of adult animals who had originated from different colonies. Occasionally, small cohorts of siblings may disperse together and be joined by a foreign animal. If this happens, all but the eventual breeding pair normally disappear from the group within about a year. This also happens in newly founded colonies where a breeder dies and is replaced by a foreign animal. The offspring of the deceased animal also usually disappear within a short time. It is unknown whether they are driven out of the colony by the new reproductive or leave of their own volition. What is certain, however, is that most, if not all, colonies in the wild are composed of the offspring of a pair of reproductive animals. Supporting data have been obtained from mitochondrial DNA analyses of entire colonies from Dordabis in Namibia. This analysis traces the maternal lineage of animals and it showed that only one male (the breeding male) in the colony is not the offspring of the reproductive female (R.L. Honeycutt, pers. comm; see also Chapter 7).

It is easy to see how a highly inbred species, such as the naked mole-rat, could sometimes produce two-queened colonies. However, in species which exhibit extreme incest avoidance and are obligatory outbreeders, as is the case in all species of *Cryptomys* studied to date, one would expect the colonies to have only one breeding female. The only way one could feasibly obtain multiple-queened colonies in the genus *Cryptomys* is if two sisters were to pair up with an unrelated and unfamiliar male and for each of the sisters to tolerate the other breeding, or for a foreign female to successfully immigrate into a colony and mate with the resident breeding male.

4.6 OVERLAP OF GENERATIONS AND LITTERS

One of the definitive features of eusocial animals is that there is an overlap of generations (Michener, 1969; Wilson, 1971). Braude (1991) and Jarvis and Bennett (1993) showed conclusively from their field data that there is an overlap of generations in wild-caught colonies of naked and Damaraland mole-rats. Indeed, this pattern appears to occur in all species of *Cryptomys* studied thus far. In addition to overlap of generations, some non-reproductives remain in their natal colony for a considerable time and directly and indirectly care for successive litters of siblings. Jarvis and Bennett (1993) found, in a study of the Damaraland mole-rat, that, of 146 animals that were under one year old on first capture, 37% remained in their natal colony for at least 16 months, 12.9% for two to two-and-a-half years and 8% for more than three years. Because litters are born to Damaraland mole-rat colonies throughout the year (approximately three a year) there is potentially an overlap of five or more litters in a colony. Jarvis and Bennett (1993) further showed, from an extensive mark and release programme, that only 8.4% of 403 animals that disappeared from their colonies successfully dispersed and found a mate.

Braude (1991), in his study in northern Kenya, has found a similar pattern of residence in colonies of naked mole-rats. In six colonies that he studied for four years, 21–80% of the animals were recaptured in the colony for a year, 15% for a maximum of two years and 2% for more than three years.

Mark–recapture studies on wild populations of social mole-rats show conclusively that recruitment to the colonies arises almost exclusively from the absorption of young into the natal colony rather than from unrelated animals joining the colony. This has been demonstrated for naked mole-rats (Braude, 1991), Damaraland mole-rats (Jarvis and Bennett, 1993) and the common mole-rat (Spinks, 1998). It is therefore probably true of all the social mole-rats that colonies consist of families of closely related individuals. Because naked mole-rats also inbreed, the degree of relatedness in this species is exceptionally high, as has been demonstrated in molecular genetic analysis. Reeve *et al.* (1990) showed, using DNA fingerprinting, that the mean coefficient of relatedness, r, amongst colony mates was 0.81, approaching that of monozygotic twins (where $r = 1.0$). Likewise, low levels of intra-colony variation have been implicated from studies of allozymes, mitochondrial DNA and major histocompatability genes (Honeycutt *et al.*, 1987, 1991b; Faulkes *et al.*, 1990a, 1997b; Chapter 7).

4.7 COOPERATIVE CARE OF THE YOUNG

Another distinguishing feature of eusocial animals is cooperative care of the young. In all colonial mole-rats studied to date, non-breeding males and females contribute to the reproductive efforts of the reproductive animals both directly and indirectly. This is an important component to reproductive success of the colony. For example, Jarvis and Bennett (1993; unpubl.) have found that the number of pups that a large, well-established colony of Damaraland mole-rats can rear is more than double that of a nascent colony with a small workforce.

All non-reproductive members of the colony contribute indirectly towards reproduction by foraging for food resources, excavating new sections of the burrow system and by defending the burrow system from conspecifics and predators. Laboratory studies on colonies of both naked and Damaraland mole-rats have shown that non-reproductives of both sexes are responsible for a large proportion of the burrow maintenance behaviour as well as the collecting and carrying of nesting material (Bennett, 1990; Lacey and Sherman, 1991). Non-reproductive *Cryptomys* carry food which they have located during burrow extension to a centrally placed store (Chapter 3). The sharing of a common food store and the relinquishing of personal reproduction would at first sight appear to be highly altruistic. However, all individuals in the colony benefit from the store and, as already mentioned, there may be considerable benefits to the non-breeder in delaying breeding while in the natal colony, so there is also a strong selfish component to these behaviours.

In terms of direct care of the young, non-breeding male and female naked mole-rats increase the time spent huddling within the nest just before the pups are born. This behaviour continues until the pups are weaned at four weeks (Jarvis, 1991; Lacey and Sherman, 1991). In the nest, the non-breeding animals frequently move, nudge and groom the pups. In contrast, in the Damaraland mole-rat, the members usually vacate the nest prior to parturition. However, once the pups are born the colony non-reproductives return to the nest and occasionally groom the pups (Bennett and Jarvis, 1988b). In both the naked and Damaraland mole-rat colonies, non-reproducing animals will retrieve pups which wander out of the nest, or that are accidentally swept along the burrow by working animals (Bennett and Jarvis, 1988b; Bennett, 1990; Lacey and Sherman, 1997) and will carry pups out of the nest in response to alarms.

As pups are weaned, another cooperative behaviour, allocoprophagy, is commonly observed. Allocoprophagy is where pups feed on the faeces of other colony members, either by begging from adults or visiting the toilet area and feeding upon the voided faeces of other animals, as in the naked mole-rat (Jarvis, 1981). The pups of the common and the Damaraland mole-rat beg faeces from their mothers after weaning (Bennett, 1990, 1992). It is an important process since the pups are consuming already semi- or completely digested food at a time when their ability to digest cellulose is poor, and also provides the pups with the endosymbiotic gut flora (caecotrophs) necessary for the digestion of cellulose (Jarvis, 1991). In several species of mole-rats, once the pups begin to wean and commence coprophagy, they become characteristically pot-bellied for approximately two months, suggesting significant changes are occurring in the gut. This condition has been observed in young pups caught in the wild and in laboratory reared individuals (Jarvis, 1991; N.C. Bennett, unpublished data), and continues in the young when they are fully weaned and for a short time during adolescence.

4.8 BEHAVIOURAL DIVISION OF LABOUR

The behavioural division of labour amongst mole-rat species has received a great deal of study and there has been much conjecture as to whether there actually is true division of labour resembling that of social insects. It was Jarvis (1981) who first described the naked mole-rat as having both a reproductive division, with 'breeding' and 'non-breeding' castes. She also described other castes, based on the frequency of burrow maintenance behaviours (nest building, digging, transporting food and transporting soil) performed by individuals, including 'frequent workers', 'infrequent workers' and 'non-workers'. The frequencies with which the animals performed burrow maintenance behaviours were further correlated with the body weight of the animal. Interestingly, frequent workers were smaller animals whose body weights were less than those of the larger infrequent workers and non-workers. Each caste was found to contain both males and females. The possession of a defined workforce may decrease the workload on the breeding female and provide her with future reproductive benefits which may not only increase her survival but also reduce the mortality of the pups born to her.

Further work by Isil (1983), Lacey and Sherman (1991), and Faulkes *et al.* (1991) also found behavioural variation amongst non-breeding animals to be strongly correlated with body weight, but were not able to identify discrete worker castes. Smaller non-breeders were found to be the most frequent participants in food carrying, soil transportation and nesting transport (Figure 4.4). Lacey and Sherman (1991) further showed by challenging colony members with foreign conspecifics or snakes, that the larger non-breeders were the primary participants in these 'defence' activities. This is of significant interest, since Brett (1991b) noted at his study site that, in the field, large, non-breeding individuals were the initial animals to visit breaches in the tunnel system. Other supporting evidence comes from Braude (1991) who suggests that it is the larger individuals that are primarily responsible for the activity of soil expulsion from the burrow, a high-risk activity known as 'volcanoing'. This is when

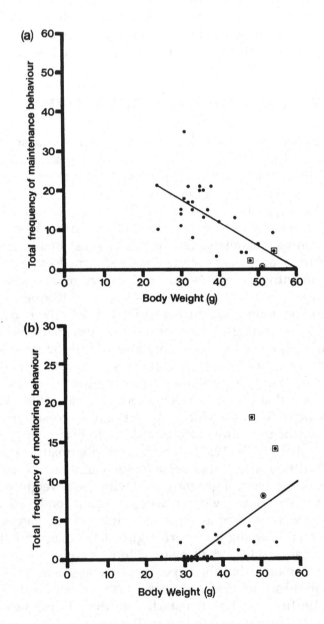

Figure 4.4. Correlation between body weight and (a) total frequency of maintenance behaviours (sweeping, digging, carrying food and nesting material, and chewing tunnel corners); $r = -0.73$; (b) total frequency of defence-related monitoring behaviours (patrolling and nest guarding); $r = 0.74$. Filled circles, non-breeding animals; filled circle within a circle, breeding female; filled circle within square, breeding male (modified from Faulkes *et al.*, 1991).

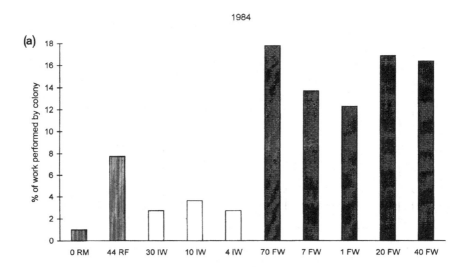

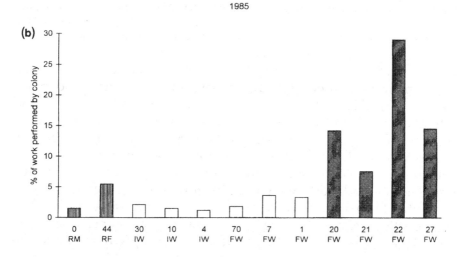

Figure 4.5. The percentage of total work performed by the reproductive and non-reproductive castes of the Damaraland mole-rat, *Cryptomys damarensis*. The non-reproductive caste is further divided into the infrequent and frequent worker groups based on the percentage of the total burrow maintenance activity performed. Animals #7, 1 and 70 move from the frequent worker group to infrequent worker group as new animals are recruited to the colony in 1985. Numbers refer to individual animals, RM, reproductive male; RF, reproductive female; IW, infrequent worker; FW, frequent worker (Bennett, 1988).

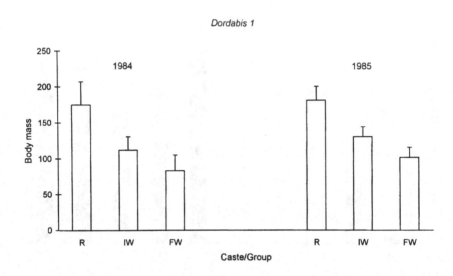

Figure 4.6. The mean body mass (g) and standard deviation of the reproductive pair, infrequent workers and frequent workers of the Dordabis 1 colony of Damaraland mole-rats. Two study periods are presented (adapted from Bennett and Jarvis, 1988b).

predation of mole-rats is most likely to take place. In contrast to the naked mole-rat, Bennett (1990), working on the Damaraland mole-rat, has been able to clearly demonstrate the presence of frequent and infrequent worker groups or castes (Bennett and Jarvis, 1988b; Jacobs *et al.*, 1991; Figure 4.5). Based on the type and amount of burrow maintenance activity, or work, individuals within a colony of the Damaraland mole-rat can be divided into two groups based on the frequency of work performed in the burrow system (Table 4.6).

In the Damaraland mole-rat, there is some support for the reproductive animals forming a morphologically distinct caste. The reproductive female usually has a distinctly elongated body and the reproductive male typically has a large, broad head. Morphologically, the non-reproductive frequent and infrequent workers can only be distinguished from one another by their body size. The frequent worker group usually consists of smaller and shorter mole-rats (Figure 4.6). Frequent workers are involved in 7.5 to 29% of the total work. In contrast, the infrequent workers are larger and heavier animals and perform less than 1.2 to 3.6% of the work (Figure 4.7). Both sexes are represented in each of the worker categories.

Table 4.6. Composition of a colony of wild-captured *Cryptomys damarensis* and the first and second litters born in captivity, showing changes in body weight and worker group from capture (August 1984) through to August 1986. The total number of bouts of work for each individual and the percentage of all work undertaken during observations are presented in parentheses (after Bennett, 1990).

Animal ID No.	Sex	Caste[a] or group	Mass at capture (g)	Work total 1984	Mass 1985 (g)	Digging & gnawing	Soil movement	Nest building	Food carrying	Total 1986
0	M	RM	198	2 (0.91)	195	1 (0.3)	4 (1.2)	0 (0)	0 (0)	5 (1.5)
44	F	RF	152	17 (7.74)	168	8 (2.4)	7 (2.1)	0 (0)	3 (0.9)	18 (5.4)
30	M	IW	134	6 (2.73)	153	2 (0.6)	3 (0.9)	0 (0)	2 (0.6)	7 (2.1)
10	F	IW	102	8 (3.64)	131	1 (0.3)	4 (1.2)	0 (0)	0 (0)	5 (1.5)
2	M	IW	142	11 (5.00)	killed	–	–	–	–	–
4	F	IW	102	6 (2.73)	115	1 (0.3)	3 (0.9)	0 (0)	0 (0)	4 (1.2)
70	F	FW[b]	102	39 (17.79)	140	1 (0.3)	4 (1.2)	0 (0)	1 (0.3)	6 (1.8)
7	M	FW[b]	78	30 (13.70)	124	3 (0.9)	9 (2.7)	0 (0)	0 (0)	12 (3.6)
1	M	FW[b]	98	27 (12.30)	120	5 (1.5)	3 (0.9)	0 (0)	3 (0.9)	11 (3.3)
40	F	FW	46	36 (16.40)	killed	–	–	–	–	–
20	F	FW	90	37 (16.90)	88	27 (8.1)	16 (4.8)	1 (0.3)	3 (0.9)	47 (14.2)
21	M	FW[c]	–	–	91	7 (2.1)	17 (5.1)	0 (0)	1 (0.3)	25 (7.5)
22	F	FW[c]	–	–	117	26 (7.8)	63 (19.0)	0 (0)	7 (2.1)	96 (29.0)
27	M	FW[c]	–	–	111	9 (2.7)	34 (10.2)	0 (0)	5 (1.5)	48 (14.5)
BB	F	FW[d]	–	–	54	6 (1.8)	4 (1.2)	1 (0.3)	2 (0.6)	13 (3.9)
GB	F	Fw[d]	–	–	52	15 (4.5)	13 (3.9)	1 (0.3)	5 (1.5)	34 (10.2)

a, caste or group: RM, reproductive male; RF, reproductive female; IW, infrequent worker; FW, frequent worker; b, changed worker group; c, first litter born in captivity; d, second litter born in captivity.

Social organisation in African mole-rats

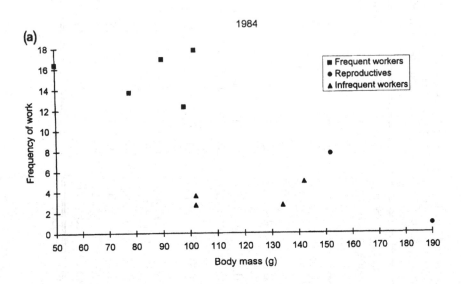

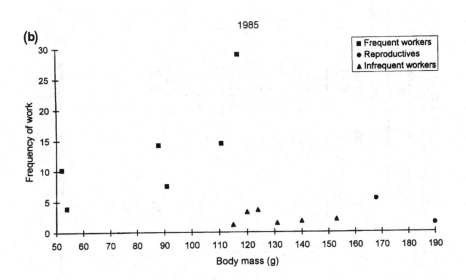

Figure 4.7. Scatter plot of the frequency of burrow maintenance behaviours and body mass for a colony of Damaraland mole-rats (*C. damarensis*) to show distinctive clusters for reproductives, infrequent workers and frequent workers (a) 1984 study period; (b) 1985 study period (Bennett, 1989).

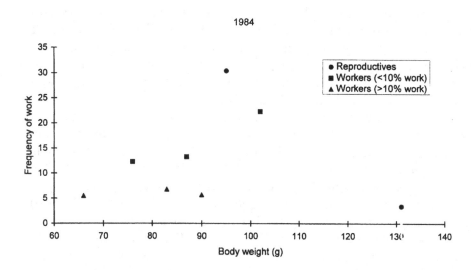

Figure 4.8. A scatter plot of the frequency of burrow maintenance behaviours and body mass of a colony of the common mole-rat (*C. h. hottentotus*) to show the lack of distinct cluster clouds between reproductives and workers (adapted from Bennett, 1989).

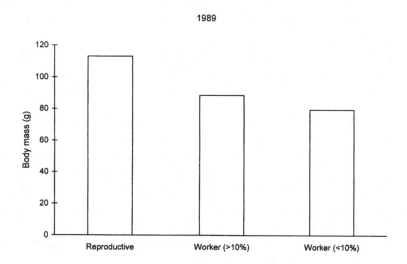

Figure 4.9. The mean body mass of the reproductive pair and non-reproductive animals in a common mole-rat colony (*C. h. hottentotus*), to illustrate that non-reproductives performing higher frequencies of burrow maintenance activities were larger than those performing < 10%. This is in direct contrast to the Damaraland mole-rat (data adapted from Bennett, 1989).

When examining the relationships between frequency of work and body mass in the common mole-rat, no relationship is found among the non-reproductive workers (Figure 4.8). Also, common mole-rats performing higher frequencies of burrow maintenance activity were generally larger individuals, rather than the smaller, as in the naked and Damaraland mole-rats (Figure 4.9).

Further to the observation that behavioural role was correlated with body size, Jarvis (1981) suggested that some naked mole-rats in her colonies exhibited age polyethism (a change in behavioural role with age). She proposed that: (i) all animals that are born to the colony enter the frequent worker caste; (ii) the slower-growing mole-rats remain permanently in the frequent worker caste and finally; (iii) the faster-growing mole-rats may become either infrequent workers or non-workers or even potential breeders. This preliminary study aroused the imagination of many sociobiologists because of the similarity to eusocial invertebrates. Further supporting evidence for a form of age polyethism came when Bennett and Jarvis (1988b) reported a similar scenario in colonies of the Damaraland mole-rat. This paper showed that: (i) most young recruits to the natal colony entered the frequent worker caste; (ii) the slower-growing adults with a small body mass remained permanently in the frequent worker caste; (iii) the faster-growing individuals became promoted to the infrequent worker caste (Figure 4.10).

Bennett and Navarro (1997) investigated the growth rates of first and second litters of pups born to pairs of animals and compared these with subsequent litters. They found that mole-rats born to the first and second litters had significantly higher absolute growth rates and reach a higher asymptote in growth than subsequent litters. Bennett and Navarro (1997) went on to further suggest that litters one and two initially became frequent workers, but with the subsequent recruitment of successive litters they attained infrequent worker or defender status. The mole-rats born to litters three and four generally showed a lower absolute body mass and became the frequent workers of the colony. Whether the first and second litters actually retard the growth of successive litters in some way, possibly by dominance interactions, is unclear and open to conjecture.

The straightforward age polyethism seen in social insects is therefore confounded in mole-rats by the fact that, at least in naked and Damaraland mole-rats, age and body weight do not always covary in a linear fashion, and behavioural role appears to be functionally related to body size rather than age *per se*. O'Riain (1996) concluded that the important distinction between mole-rat colonies

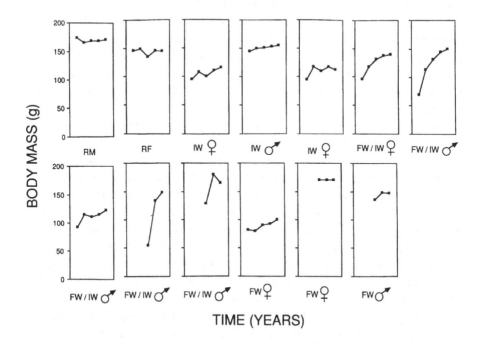

Figure 4.10. Graphs of the mean body mass of individuals from a Damaraland mole-rat (*C. damarensis*) colony taken at six monthly intervals for a period of two-and-a-half years to show that animals which moved from a frequent worker grouping to infrequent worker grouping (FW/IW) showed a dramatic change in body mass. In contrast those animals remaining in their groups (infrequent worker, IF and frequent worker, FW) over the entire period, showed no such change in body mass (RF, reproductive female; RM, reproductive male) (data from Bennett, unpubl.).

and insect societies is that mole-rats do not perform different roles, they only perform different amounts of the same task (increased or decreased frequencies). Thus, in mole-rats, task repertoires of all non-breeding mole-rats are in essence the same, the differences merely being quantitative. In contrast, in eusocial insects the tasks are qualitative and furthermore, morphological specialisation of castes also occurs.

In social insect societies, extensive and irreversible morphological specialisation is often accompanied by sterility in the organism. In mole-rats, the retention of a functional gonad (although in females, the ovaries are sometimes prepubescent) makes it possible for that individual to become sexually active and produce offspring should the opportunity arise. Hence, by showing only a quantitative behavioural task specialisation, the mole-rat retains a degree of plasticity and can switch either the behavioural task or the frequencies of particular tasks. This seems a logical outcome when one considers the dynamic nature of colonies in many species of mole-rats. Also morphological specialisation in vertebrates may be constrained by their patterns of development. In insect societies development is short and certain individuals can be selectively modified by pheromonal messages (like royal jelly in bees to produce new queens) to give rise to morphs suited to particular specialist tasks (Wilson, 1971).

The naked mole-rat may be one of the few exceptions to the rule, with some evidence for morphological specialisation. The breeding queen has a distinctly elongated body, which X-ray measurement has shown to be due to lengthening of the vertebrae (Jarvis *et al.*, 1991). O'Riain (1996) reported the presence of defender mole-rats in his colonies that were characterised by certain phenotypes. There is evidence of polymorphism among non-breeders for incisor width (the weapons of defence) as well as for jaw musculature. O'Riain (1996) observed that colony defenders were typically characterised by hypertrophied temporal musculature which resulted in these animals exhibiting a typically 'square-shaped head'. O'Riain *et al.* (1996) also announced the finding of a disperser morph that was typically large and contained extensive fat reserves. Moreover, measurements of the basal circulating plasma luteinising hormone in these animals revealed a profile more reminiscent of a breeding animal than that of a non-breeder (Chapter 5).

4.9 DOMINANCE AND HIERARCHIES IN SOCIAL MOLE-RATS

An inevitable consequence of living in a group is that, due to variation in behaviour, certain individuals are more 'assertive' than others and, through agonistic interactions, may monopolise resources. The phenomenon of the dominance hierarchy was first investigated by Schjelderup-Ebbe (1922), who worked on chickens, and from where the phrase 'pecking order', describing the dominance rank position, derives. A social hierarchy presumes that some members of the group will seek authority by fighting or by some other special achievement, whilst low-ranking individuals will accept this order. Dominance hierarchies may be linear (in which there is a discrete peck order) or non-linear (in which several individuals may be codominant with one another). In a wide variety of social organisms, dominance orders are established through agonistic (aggressive) interactions between members and hence usually reflect the competitive abilities of individuals, with an individual's fitness being closely linked to its rank within the group (Dewsbury, 1982).

Figure 4.11. A dominant breeding queen naked mole-rat passes over a subordinate non-breeder, and shoves a second subordinate colony member. Both of these behaviours are signs of dominance.

In social mole-rats, dominance interactions may play a critical role in the establishment and maintenance of social order. This is particularly an issue with naked mole-rats, where, due to inbreeding and the replacement of reproductives from within the colony, there is competition for the attainment of reproductive status. In order to investigate dominance, agonistic behaviours and 'winners' of agonistic interactions between individuals need to be quantified. This is not always a simple task as, in established groups, overt aggression is often replaced by more subtle cues that are sufficient to maintain a hierarchy, but make behavioural quantification difficult. Thus, in naked mole-rats, there is normally little in the way of aggressive encounters, with the exception of shoving by the queen (Figure 4.11), and there are contrasting reports about the type of dominance hierarchy that is present.

Clarke and Faulkes (1997) showed that there was a linear dominance hierarchy in naked mole-rat colonies. This was determined by analysing both the relatively rare agonistic interactions, and a behaviour that casual observations suggested reflected dominance, and was commonly observed in all the colony members, namely the asymmetry of passing behaviour in pairs of animals meeting in tunnels (Figure 4.11). When two naked mole-rats meet in a tunnel face-to-face, after a period of brief mutual sniffing of the facial area (implying recognition) one animal, presumed to be the dominant one of the dyad, passes over the top of the other (except for rare side-to-side passes). Dominance rank, calculated from this passing behaviour, showed a strong, and highly significant, correlation with dominance rank calculated using agonistic behaviours ($r_s = 0.85$, $p < 0.001$). Using these calculations, the breeding female is almost always found to be the highest-ranking colony member, and dominance positions for entire colonies have been calculated.

Clarke and Faulkes (1997) found that in three colonies there was a highly significant negative correlation between mean body mass and dominance rank ($r_s = -0.91$, $p < 0.001$; $r_s = -0.85$, $p < 0.001$ and $r_s = -0.90$, $p < 0.001$ respectively). Dominance rank position calculated from passing behaviour was also negatively correlated with age and urinary testosterone levels. Thus, the most dominant animals within the colony tended to be the oldest, largest, and those having the highest testosterone titres (Figure 4.12). All of these variables have been implicated in the attainment of high dominance rank in other mammals (Clutton-Brock, 1988). These high-ranking males and females in naked mole-rat colonies were also the most likely animals to become replacement breeders on removal of the existing queen or breeding males (Clarke and Faulkes, 1997, 1998).

Interestingly, Schieffelin and Sherman (1995) also found a dominance hierarchy in naked mole-rat colonies based on body weight but not age or sex, using tugging contests over food to determine dominance relationships. Jarvis *et al.* (1991) and O'Riain (1996) have suggested that body weight is labile and does not covary with age, possibly explaining why, in the study by Schieffelin and Sherman, dominance rank was correlated with weight but not age. However, O'Riain (1996) has suggested that mole-rats do not exhibit a linear dominance pattern. He suggests that the dominance structure amongst naked mole-rats typically consists of the single breeding female, who is dominant to a number of subordinate individuals, with a degree of circular triads, which results in a low linearity

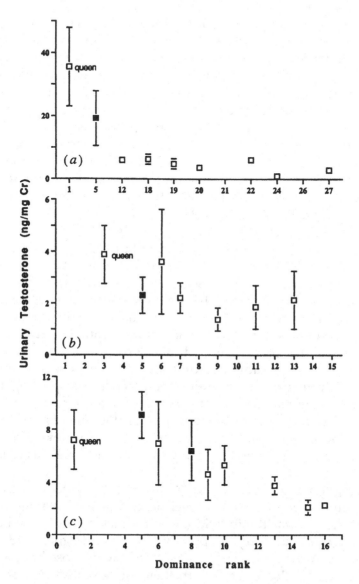

Figure 4.12. Urinary testosterone levels (mean ± S.E.M.) as a function of female dominance rank in three naked mole-rat colonies. Filled squares indicate those females which were to succeed as new queens when the original queens were removed from the colony. Spearman rank correlation coefficient values (r_s) and levels of significance were as follows: (a) $r_s = -0.89$, $p < 0.01$; (b) $r_s = -0.68$, n.s.; (c) $r_s = -0.95$, $p < 0.01$; (reproduced from Clarke and Faulkes, 1997 with permission from *Proceedings: Biological Sciences*).

coefficient. The different types of hierarchy recorded by Clarke and Faulkes (1997), and O'Riain (1996) may be due to different behaviours being used to assess dominance.

In *Cryptomys*, a range of agonistic interactions is more commonly observed than in naked mole-rats, and as colony sizes are also smaller, dominance hierarchies are easier to quantify. In all species of Bathyergidae, 'sparring' appears to play an important role in the development of the pups, possibly functioning to establish dominance positions early on in the life of an animal. In *Cryptomys*, sparring begins at about 10 days after birth (Bennett and Jarvis, 1988b; Bennett, 1990), while naked mole-rat pups begin to spar at 14 to 20 days old (J.U.M. Jarvis, pers. comm.). The solitary Cape dune mole-rat and Namaqua dune mole-rat begin to spar 13 and 16 days after birth (Bennett *et al.*, 1991), whilst the Cape mole-rat begins later, at 35 days (Bennett and Jarvis, 1988a).

In all the solitary genera, sparring occurs between litter mates and the intensity of this behaviour increases until injury is inflicted or the pups disperse. It is possible that in the young mole-rats the increased frequency of sparring is a consequence of play. However, as the solitary pups develop, sparring becomes more intense with escalation into fighting. This does not take place in the social mole-rats. In the social bathyergids, sparring is initially almost entirely between litter mates but, as the pups age and develop, this behaviour occurs between juveniles and adults. Adult animals are extremely tolerant towards the young pups and allow the juveniles to spar vigorously with them. In the naked mole-rat, sparring (also referred to as tooth fencing and incisor tussles) continues until the animals are around two years old, then it suddenly becomes infrequent. Likewise, adult cryptomids do not spar as frequently as pups and juveniles (Bennett, unpubl.).

Sparring can occur between adults, siblings or adults and siblings. The two animals gently lock their incisors and then have a tug of war. During the tug of war the individuals brace themselves in the burrow with the forefeet and try to pull one another (Figure 4.13). The episodes of pulling may be interspersed with nose-butting and pushing at each other's faces with their forefeet. The skin is never bitten and the interactions are usually terminated either by one animal rolling over on to its back and exposing its belly and genitalia or by one animal reversing at high speed along the tunnel system. Adult–sibling and sibling–sibling interactions are probably important in determining the dominance status of individuals in the colony. Adult–adult interactions are infrequent and probably more crucial for the maintenance of dominance positions or in challenging dominance position.

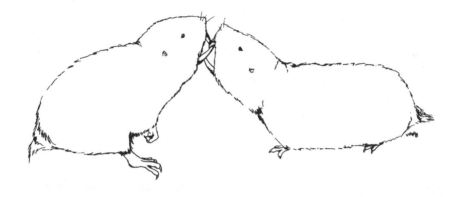

Figure 4.13. Sparring is characterised by two animals gently locking their incisors and then bracing themselves against one another. The intensity of sparring varies, but wounds are not normally inflicted. Episodes of pulling are interspersed with nose-butting and pushing at one another's faces.

Another agonistic interaction, tail-pulling, occurs in all the social mole-rats during cooperative burrowing (Jarvis, 1969a; Bennett, 1988). In the genus *Cryptomys* it appears to be associated with the dominance of particular individuals, the dominant mole-rat pulling the tail of the subordinate.

When two animals are moving in the same direction within the burrow, it is not uncommon for the anterior mole-rat to stop and feed or rest. The posterior mole-rat, in turn, may pull the obstructing individual back along the burrow by its tail (Figure 4.14). The animal being towed usually turns onto its side and then its back and attempts to brace itself against the sides of the burrow. Once released, the mole-rat will often return to the place from which it was initially towed and tail-pulling may be initiated all over again.

Other agonistic behaviours include alarm and threat behaviours used in response to disturbances from outside the colony. These are head-back threat posture, roll-over threat and pumping. In the head-back threat posture, the animal postures with its head thrown back, its eyes open and its mouth fully agape. The forefeet are placed firmly in front of the mole-rat and the hind limbs are widely spaced laterally but braced for a rapid advance or retreat. Periodically, the mole-rat snorts and chatters its teeth. The posture is firm and rigid

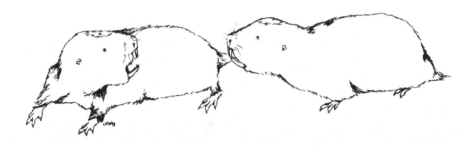

Figure 4.14. Tail-pulling is used to displace one animal from its digging position or resting position. The obstructing individual is usually pulled along the burrow by its tail.

Figure 4.15. Head-back threat posture. The head is thrown back, exposing the dagger-like teeth.

with the animal making short jerks or jumps towards the agonistic source (Figure 4.15).

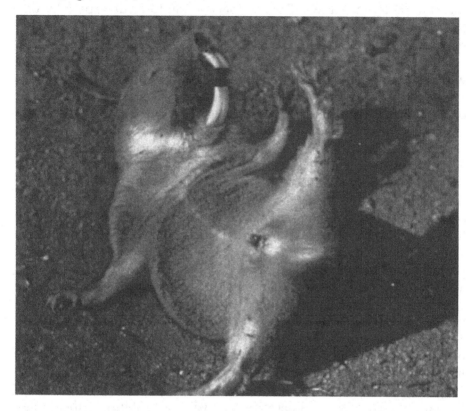

Figure 4.16. Roll-over posture. The mole-rat rolls onto its back, keeping its body curved. The head is lifted and the mouth is wide agape (photo © Tim Jackson).

The roll-over threat posture has been observed in both the Damaraland and Mashona mole-rats (Bennett, 1990; pers. obs.), and is characteristic of these species. The mole-rat rolls onto its back, keeps its body curved and supports itself with its forepaws placed behind or lateral to the body. Its head is lifted and the mouth is held widely agape. The tail is tucked into the pelvic region. The mole-rat's belly and genitalia are exposed, but are largely protected by the prominent incisors framing the gaping mouth. Vocalisations are emitted in the form of grunts, snarls and twitters (Figure 4.16).

The final alarm behaviour, pumping, occurs when the mole-rat cautiously approaches the source of the threat, repeatedly sniffing the air and holding its tail out straight. While still 10 to 20 cm or more away, the animal postures, with its fore and hind legs splayed

laterally, its head stretched out forward and its body flattened
dorso-ventrally. The mole-rat then pumps its hind region (legs and
sacral area) up and down. This action is quite forceful, especially on
the downstroke. In a very high-level threat situation the whole body
may be lifted off the ground with the upstroke.

Because a greater variety of agonistic behaviours can be recorded
with comparatively high frequencies, determination of dominance
rank and the assessment of hierarchy linearity in many of the species
of *Cryptomys* has been estimated using different methods to those
employed in naked mole-rat studies. In essence, the approach, first
reported by Aspey (1977), uses an objective, computational method,
and also a more subjective method involving the construction of a
dominance matrix. The results obtained from these two methods are
then compared for their level of similarity. In the former, a multi-
correlation of the frequency of interactions is used to reduce the
interactive behaviours of the animals into a few discrete factors.
These factors normally group together animals showing similar
patterns of behavioural interaction. Likewise, in the subjective
approach to dominance determination, a dominance index is
constructed for each animal and the interactions with every other
individual in the colony are used to construct a dominance matrix.
The dominance index for a mole-rat is calculated using the sum total
of several parameters. These are the behavioural type, its subjective
intensity weighting (i.e whether aggressive or passive), and the
frequency with which that interaction occurs between individuals in
the colony. The use of subjective intensity weightings is the same
whether it be sparring in the Damaraland mole-rat or the Mashona
mole-rat. In this sense uniformity is maintained in the calculation of
the dominance index. Using the dominance indices of all dyadic
encounters between particular pairs of mole-rats, the number of
individuals over which each animal is dominant to, or subordinate to,
can be calculated.

The dominance relationships evident within Damaraland mole-rat
colonies, in which the reproductive animals are positioned at the
apex of the hierarchy, are similar to those in the naked mole-rat. In
three colonies of the Damaraland mole-rat, the dominance hierarchy
was found to be linear (Landau's index, $h = 0.9$ to 1.00; Bennett,
1988; Jacobs *et al.*, 1991). Without exception, the reproductive male
was at the top of the hierarchy with the reproductive pair being the
most dominant members of their respective genders, followed by non-
reproductive males (Bennett, 1988; Jacobs *et al.*, 1991). There appears
to be a slight trend towards less linearity in the hierarchy in the less

social species (Table 4.7). In the common mole-rat, where colony size is smaller (mean = 5), Landau's index ranges from 0.56 to 0.85 although, again, the reproductive male is at the apex of the hierarchy (Bennett, 1989; Rosenthal *et al.*, 1992). Like the Damaraland mole-rat, the reproductive pair are the most dominant individuals of each respective gender and non-reproductive males are more dominant than non-reproductive females. This trend of the reproductives being at the top of the hierarchy has been also been found in the Mashona mole-rat, the giant Zambian mole-rat and the highveld mole-rat (*C. h. pretoriaë*) (Gabathuler *et al.*, 1996; Wallace and Bennett, 1998; Moolman *et al.*, 1998).

Table 4.7. Comparison of factors related to sociality in five bathyergid species, to illustrate the 'eusociality continuum'. Landau's index expresses dominance hierarchies on a linear scale from zero (non-linear) to one (linear). Adapted from Wallace and Bennett, 1998.

Species	Habitat	Colony biomass (kg)	Mean colony size	Landau's index	Social status
H. glaber	arid	2.46	70–80	1.0	eusocial
C. damarensis	arid	2.32	18	0.94–1.0	eusocial
C. mechowi	mesic	2.34	9	0.83	social
C. darlingi	mesic	0.45	7	0.77	social
C. h. hottentotus	mesic	0.66	14	0.56–0.85	social

4.10 SOME CONCLUDING REMARKS

The Bathyergidae is the only subterranean rodent family with highly social species. A number of authors have proposed that the distribution, size and digestibility of the geophytes, as well as the variation and predictability of the rainfall, have played a crucial role in shaping the sociality that has arisen within the Bathyergidae (Jarvis, 1978; Bennett, 1988; Lovegrove and Wissel, 1988; Lovegrove, 1991; Jarvis and Bennett, 1993; Jarvis *et al.*, 1994; Faulkes *et al.*, 1997a). The importance of these ecological factors has given rise to the 'food-aridity hypothesis'. The basis of this hypothesis is essentially threefold, namely (i) that geophytes are larger in size but occur at lower densities in more arid regions (Brett, 1986, 1991a; Bennett, 1988; Jarvis and Bennett, 1990); (ii) mole-rats forage in a 'blind fashion' for these geophytes (Jarvis *et al.*, 1998), so that in arid areas the chances of finding food resources are increased, and the risks of unsuccessful foraging reduced, when large numbers of animals are involved in food searching, and (iii) because the energetic costs of burrowing are greater when the soil is compact and dry (Vleck, 1979; 1981), mole-rats limit most of their food-related digging to times when the rain has moistened the soil (Jarvis and Bennett, 1993; Jarvis *et al.*, 1994; 1998). It has been conclusively shown that these ecological constraints have given rise to a trend of increasing group size and the level of social organisation with a decreasing geophyte density and rainfall (Jarvis *et al.*, 1994; Faulkes *et al.*, 1997a). Whilst these two factors preclude solitary species from arid regions, they do not prevent social species from invading the more mesic environments. So why has sociality not arisen in any of the other subterranean families and sub-families to the degree that it has in the Bathyergidae? Lacey and Sherman (1997) have suggested that if the food–aridity hypothesis holds true across all subterranean rodent taxa, then it would be expected that sociality should not arise in regions with (i) food evenly distributed throughout the environment; (ii) where burrowing is energetically inexpensive and burrowing is feasible for much of the year, and (iii) The seasonality of the environment should not be too marked (i.e. rainfall should not be sporadic and unpredictable).

Clearly, there must be other factors involved, since the spalacids also occupy an aridity gradient, in which (i) the food resource is clumped (Galil, 1967); (ii) burrowing is energetically very expensive (Arieli *et al.*, 1977), and (iii) there is a marked seasonality to the environment (Shanas *et al.*, 1995). Mole-rats in the genus *Spalax* are all strictly solitary. However, subspecies of *Spalax* occurring in more arid habitats tend to have lower levels of aggression and are more pacifistic than their mesic counterparts (Ganem and Nevo, 1996). Ganem and Nevo (1996) found that in *Spalax* occurring in different zones of aridity, higher levels of aggression may be selected against in the more xeric habitats due to the physiological correlates (increased corticosterone levels and associated decreased water economy). In the less arid dwelling subspecies of *Spalax*, there were increased levels of corticosterone and a decreased water economy. The phylogeny of the Bathyergidae may be important because sociality in the subterranean Bathyergidae is unique to this family. The bathyergids also have strong hystricomorph affinities (Jarvis and Bennett, 1990). The predominance of social species in the hystricomorphs may have influenced the development of sociality in this group (Burda, 1990). It remains an interesting enigma that sociality has evolved to an advanced degree in these subterranean rodents, since the general rule in subterranean rodents and insectivores is to be strictly solitary and highly xenophobic (Nevo, 1979). In the African mole-rats, it is not mere coloniality that has arisen but the formal sharing of a unified burrow system which is maintained and defended by all members of the colony.

Assuming that the ancestral form was solitary, how could sociality have arisen? Jarvis *et al.* (1994) have suggested that sociality may have arisen as a result of prolonged droughts which, through evolutionary history, may have resulted in ecological 'bottlenecks' serving to exclude solitary genera from arid areas. It is envisaged that only those species in which the young did not disperse at weaning, but rather remained at the natal burrow, were able to survive in such an unpredictable environment. The dichotomy in habitat selection exhibited at present thus appears to be the result of selection for philopatry in particular species. Thus, solitary species today occur in mesic areas which are characterised by higher densities of geophytes and rainfall patterns which are regular and predictable. In contrast, social species tend to occur in semi-arid and arid regions characterised by lower geophytic densities and more sporadic and unpredictable rainfall. The bathyergids thus provide us with an excellent research tool with which to further our knowledge of the evolution of sociality in mammals, particularly those inhabiting underground nests.

Chapter 5

Life history patterns and reproductive biology

5.1 INTRODUCTION

In this chapter we investigate the effect that cooperative living has upon the reproductive and life history patterns of social bathyergids, in comparison to the solitary dwelling species of the Bathyergidae and taxonomically unrelated solitary subterranean rodents. Reproductive and life history parameters are notoriously difficult to study in subterranean animals, although recent developments in molecular genetic techniques should overcome some of the difficulties, particularly when determining mating strategies.

Reproductive parameters such as the gross anatomy of the reproductive tract and its physical dimensions, follicular development and pregnancy in females, and presence/absence of spermatozoa in males, have been obtained in five of the families of subterranean rodents through examination of specimens collected at different times of the year. This gives a kind of 'snapshot' picture of events (Jarvis, 1969a,b, 1973b; Van der Horst, 1972; Smolen et al., 1980; Redi et al., 1986; Bennett and Jarvis, 1988b; Flynn, 1990; Reig et al., 1990; Malizia and Busch, 1991; Simson et al., 1993). Longitudinal studies which follow individuals over time, recording life history parameters such as courtship, copulation, gestation, parturition and pup development, are considerably more sparse (see Bennett et al., 1991, for review) but have been attempted for a number of species in the family Bathyergidae (Jarvis, 1969b; Jarvis and Sale, 1971; Bennett and Jarvis, 1988a,b; Bennett, 1989; Faulkes et al., 1990a Bennett et al., 1994a,b; Bennett and Aguilar 1995).

127

5.2 ENVIRONMENTAL INFLUENCES ON REPRODUCTION

Post mortem examination of specimens collected at different times of the year indicates that many solitary subterranean rodents exhibit a seasonal breeding pattern, for example, species in the families Ctenomyidae (Reig *et al.*, 1990; Malizia and Busch, 1991), the Geomyidae (Smolen *et al.*, 1980), the Rhizomyidae (Jarvis, 1973a,b; Flynn, 1990) the Bathyergidae (Jarvis, 1969b; Van der Horst, 1972; Bennett and Jarvis, 1988a) and the Spalacidae (Redi *et al.*, 1986; Simson *et al.*, 1993). As well as social cues, environmental factors may have a profound influence on reproductive function, although there is relatively limited information available on exactly what the environmental factors are that control reproduction, and the timing of breeding, in subterranean rodents that reproduce seasonally. In species that are exclusively subterranean with small, vestigial eyes and a degenerate visual system, one might expect that light (photoperiodic cues) would not play a major role in controlling reproduction. However, it has been shown that even the blind mole-rat, *Spalax ehrenbergi*, is able to entrain daily activity patterns to a circadian light–dark cycle, despite having vestigial eyes that are totally subcutaneous (Cooper *et al.*, 1993). Thus, light as a cue should not be totally dismissed as a potentially important environmental zeitgeber even in animals that are poorly sighted or effectively blind. Intuitively it might also appear that temperature is not an important cue, since temperatures in the primary burrow system deeper than 30 cm are relatively constant compared with those occurring on the soil surface (Gates 1962; see also Figure 2.2, Chapter 2). However, because the majority of the burrow system of a subterranean rodent may be composed of more superficial 'foraging' tunnels, built in search of roots and tubers, animals entering these tunnels are exposed to both diurnal and seasonal fluctuations in temperature. Thus, despite the constant temperatures deep in the burrow, changes in temperature in the superficial tunnels could act as a stimulus. Many of the solitary subterranean rodents occurring in the higher latitudes exhibit a marked breeding season and the environmental cues triggering the onset of breeding may well be thermal, resulting from annual changes in burrow temperature (Bennett *et al.*, 1988). Seasonal rainfall is, potentially, also a strong proximate cue resulting in the moistening of soil, thus prompting the extension of existing sections of the burrow and facilitating colony fission and dispersal. Indeed, three species of bathyergid mole-rats occurring in the Cape Province of South Africa (the dune mole-rat, *Bathyergus suillus*, the Namaqua dune mole-rat, *B. janetta*, and the Cape mole-rat, *Georychus capensis*), have adapted their breeding season so that the juveniles develop and establish their own burrow systems in early spring when the rainfall enables the soil to be easily worked. Many of the tropical mole-rats occurring in the lower latitudes lack a breeding season and this probably is due to an absence of environmental triggers and the thermally stable environment of the burrow.

5.3 COURTSHIP, MATING AND OVULATION

The restriction of a single animal to a burrow system, which is the norm for most subterranean rodents, means that for males and females to come together to mate, the strong barrier of xenophobia and mutual aggression must be tempered. Whatever the taxonomic group, initial communication and attraction appear to be mediated via seismic signalling. As the name suggests, these signals are vibrations which are carried through the substrate and function to advertise the sex, status and intention to breed of the propagator. Such seismic signals are known to be transmitted through soil at levels an order of magnitude greater than auditory signals (Narins *et al.*, 1992). Seismic communication by hind foot drumming has been reported in the family Geomyidae (O.J. Reichman, pers. comm.) and in several species in the family Bathyergidae, including the Cape mole-rat, the Cape dune mole-rat, the Namaqua dune mole-rat, the common mole-rat and the Damaraland mole-rat (Bennett and Jarvis, 1988a,b; Jarvis and Bennett, 1990, 1991).

In the Cape mole-rat, hind foot drumming observed in captive colonies is initiated by the male, who drums by striking his hind feet on the floor. This drumming is extremely fast, each pulse being 0.035 seconds in duration, compared with the slower response of the female at 0.05 seconds per pulse. In the field, drumming can even be heard above the surface for a distance of 10 m from the source (Bennett and Jarvis, 1988a; Figure 5.1). Thus, seismic communication in this species is almost living up to its literal meaning, that is, belonging to or produced by an earthquake!

Another form of seismic sound propagation is incisor tapping, a behaviour which has been observed in the rhizomyid, *Tachyoryctes splendens* (Jarvis, 1969a). In this species the upper incisors are rapped on the base of the burrow, thus announcing the presence to a potential mate.

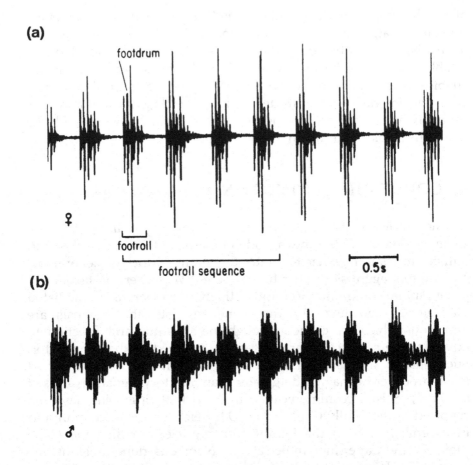

Figure 5.1. Simultaneously-recorded auditory waveforms of the 'thumps' produced during a footroll sequence by a female (a) and a male (b) *G. capensis* mole-rat. The animals were separated by a transparent divider located in a plexiglass tunnel system through which the mole-rats could travel. Note the 1:1 synchrony of the footrolls throughout the interaction (reproduced with permission from *J. Comp. Physiol.*; Narins *et al.*, 1992).

In the spalacid, *Spalax ehrenbergi*, head drumming appears to be the principal means of communication, with the signalling mole-rat tapping the upper part of its head against the ceiling of the tunnel (Heth *et al.*, 1987; Rado *et al.*, 1987). The vibrations are transmitted through the soil and sensed by the receiver which uses its lower jaw to detect the seismic communication (Rado *et al.*, 1989). There is currently no record of seismic communication in *Ctenomys* (Altuna *et al.*, 1991). Because the social species of bathyergids generally communicate within the confines of a burrow, rather than between burrows, seismic signals are possibly less important than in the solitary dwelling species. Instead, tactile, vocal and olfactory cues are utilised in courtship and mating. Of course, these systems of communication are not mutually exclusive and seismic communication may be important in dispersal in social mole-rats. Tactile, vocal, and olfactory cues are also important in solitary mole-rats once they have attracted each other with seismic signals.

Once a reproductive pair have bonded, the copulatory behaviours of a number of taxonomically unrelated solitary subterranean rodents are very similar. Many of the species studied show no lock during copulation, and exhibit thrusting during intromission, multiple intromission and multiple ejaculation, the so called 'Pattern 9' from Dewsbury's copulatory categorisation (Dewsbury, 1975). In *S. ehrenbergi* (Nevo, 1961), *Tachyoryctes splendens* (Jarvis, 1969b), *G. capensis* (Bennett and Jarvis, 1988a), *C. pearsoni* (Altuna *et al.*, 1991) and species of *Thomomys* (Andersen, 1978) and *Geomys* (Schramm, 1961), courtship is usually initiated by the male.

In all documented cases of mating described for subterranean rodents there are repeated mounting sequences, interspersed by short periods during which the animals are involved in bouts of grooming, particularly around the genitalia. Andersen (1978) suggested that the extended copulatory sequence generally found in *S. ehrenbergi*, *Thomomys talpoides* and *T. bottae* (blind mole-rats and pocket gophers respectively) may reflect the relative safety provided by the burrow in underground breeders, and the protective confines of the burrow could certainly promote extended courtship and copulation.

We have mentioned in Section 4.3 that sexual dimorphism in body size in the solitary Cape dune mole-rat and Namaqua dune mole-rat is extremely marked (Taylor *et al.*, 1985; Jarvis and Bennett, 1991). Reichman *et al.* (1982) suggest that in the geomyid *T. bottae* only the largest and most aggressive males mate and these are polygynous, mating with a number of females. It is possible that a similar mating strategy may occur in the two sexually dimorphic *Bathyergus* species. Indeed, the Cape dune mole-rat males are characterised by a thick pad of skin (1 cm thick) on the ventral surface of the neck (Davies and Jarvis, 1986) which may be used as a means of protection during fights. Badly injured Cape dune mole-rat males have been found in the field with broken incisors and deep wounds around the head, indicative of incisor punctures from a conspecific (Jarvis and Bennett, 1991). Additional data supportive of male–male aggression and competition comes from the finding of two interlocking skulls of male Namaqua dune mole-rats, in which the incisors of one mole-rat had breached the nasal bone of the other. Evidently the two combatants had died being, literally, locked together (Figure 5.2; J. du G. Harrison, pers. comm.). It is not known why the solitary Cape mole-rat exhibits no marked sexual dimorphism, since male–male aggresssion is expected to be just as high as in Cape dune and Namaqua dune mole-rats. There is sexual dimorphism in body size in some of the social bathyergids but the situation is complicated by the fact that among the non-breeding animals, body size is also related to social status (Jarvis *et al.*, 1991; O'Riain, 1996; Clarke and Faulkes, 1997). In the cryptomids, the breeding male tends to be larger than the breeding female, whereas in naked mole-rats the queen is usually the largest individual within the colony, with a distinctive elongated body (Jarvis, 1991; O'Riain, 1996).

Figure 5.2. The skulls of two male *Bathyergus janetta*, who apparently died in mortal combat after the incisors of one became lodged in the skull of its adversary during a fight (photograph Terry Dennett).

Because of their aggressive nature, solitary species of mole-rat are extremely difficult to breed in captivity. However, mating in the Cape mole-rat has been observed in the laboratory. Copulation is brief, involving multiple, brief intromissions of 2–3 thrusts per second, during which the female is completely docile and adopts the lordosis

Figure 5.3. The behavioural sequence of events leading to mating in the Cape mole-rat, *Georychus capensis*. (a) The sexually active male and female meet head to head after the elaborate seismic communication which attracts them together and helps to break down the normal xenophobic reaction. (b) The male approaches the female from the rear, producing a gutteral twitter as he comes close: the female is very wary at this time; (c) the male mounts the female and grips her with his hind feet. (d) The male pushes the hind quarters of the female with his hind feet and grips the shoulders of the female with his forepaws. (e) The female adopts lordosis, the male thrusts and intromission occurs.

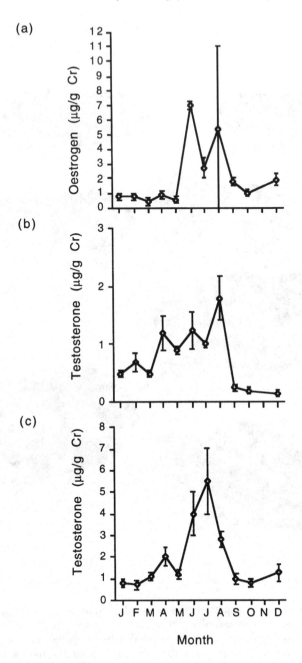

Figure 5.4. Mean monthly concentrations (± S.E.M.) of (a) urinary oestrogen in the female; (b) urinary testosterone in the female, and (c) urinary testosterone in the male of a breeding pair of captive Cape mole-rats (*Georychus capensis*). Values are expressed as µg of hormone per mg of creatinine (Cr).

posture characteristic of rodents (Figure 5.3). Pelvic thrusting becomes more frequent prior to ejaculation (Bennett and Jarvis, 1988a). The female chases the male away after, a behaviour also recorded in the rhizomyid *T. splendens*. The strong seasonality in breeding in the Cape mole-rat is reflected in hormone profiles obtained from serially collected urine samples from males and females maintained in captivity throughout the year. Both males and females show peaks in testosterone and oestrogen between May and September, the time when courtship, copulation and subsequent pregnancies are observed in the field (Bennett and Jarvis, 1988b; Figure 5.4).

Courtship in the social bathyergids is generally more involved than in the solitary species. In the common mole-rat, courtship is initiated by the male when he encounters a receptive oestrous reproductive female and begins with both animals vocalising, the male twittering and the female chirping. The female then raises her tail exposing the genital region, which is in turn smelt by the male. The male takes the female's rump into his mouth and chews gently. The male also occasionally strokes the female's side with his head. After this more elaborate courtship, the male mounts the female, restraining her by biting at the back of the neck and thrusting. After a loud squeal, the pair disengage and groom their respective genitalia (Bennett, 1989).

In the Damaraland mole-rat, it is the female that initiates courtship (Figure 5.5). In studies on captive colonies, on encountering the reproductive male in the burrow, the reproductive female vocalises, briefly drums using the hind feet and mounts the head of the male. The pair usually enter a chamber and chase one another in a head-to-tail fashion in a tight cycle. The female pauses, raises the tail and adopts the lordosis posture. The male smells the genitalia of the female, mounts and mates. There is no post-mating chasing (Figure 5.5; Bennett and Jarvis, 1988b; Bennett, 1990). The mating sequence occurs frequently for about 10 days and then ceases, presumably when either the female conceives or oestrus ceases.

In the naked mole-rat the female also initiates courtship and mating. Usually she vocalises and then crouches in front of a breeding male, adopting the lordosis position, whereupon the male mounts (Jarvis, 1991). On occasions mutual ano-genital nuzzling and sniffing may also occur prior to or during courtship (Lacey *et al.*, 1991). Mating is usually confined to a few hours and occurs around 10 days post-partum (Jarvis, 1991). In colonies that contain more than one breeding male, the breeding queen may mate with them all during any particular period of oestrus (Jarvis, 1991).

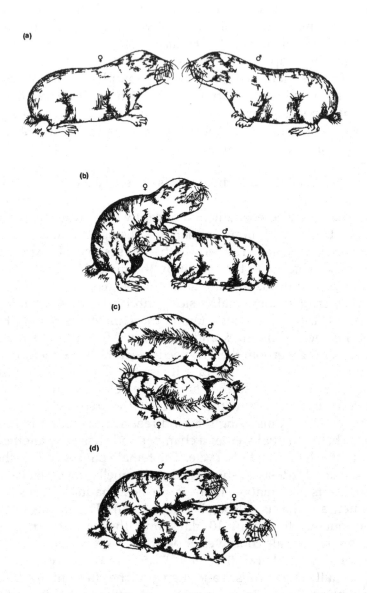

Figure 5.5. The behavioural sequence of events leading to mating in the Damaraland mole-rat, *Cryptomys damarensis*. (a) The male is solicited by the female in a head-to-head confrontation. (b) The female head-mounts the male and raps her hind feet on the burrow floor. (c) The male and female pursue one another in a head-to-tail chase, with the female's tail raised. (d) The male seizes the female by her tail using his incisors. He mounts, holding her around the shoulders with his forelimbs, and bites her neck. The female responds by raising her head and flattening her body, and copulation ensues.

Breeding male naked mole-rats (*Heterocephalus glaber*) normally start as comparatively large individuals from the upper ranks of the colony hierarchy (Clarke and Faulkes, 1998), but with time they consistently lose body mass and condition and can be readily identified by their emaciated state (Jarvis, 1991). It has been estimated that, on average, these breeding males lose 17 to 30% of their body mass from their heaviest weight prior to attaining 'alpha' male status. The exact cause of this wasting is unclear, but may be due to the consequences of the periods of high testosterone recorded in these males, which may lead to chronic immune suppression. Certainly, in the marsupial shrews of the *Antechinus* genus, seasonally-induced high testosterone and corticosteroid levels have been implicated in the demise of these males following their annual mating frenzy. These animals adopt a very unusual strategy in that all the males die off after the mating season, leaving the pregnant females to carry on the species. Deaths in the males appear to arise from a variety of causes, including lowered disease resistance and stress ulcers in the digestive tract.

Family analysis of a captive colony of naked mole-rats using multilocus DNA fingerprinting has revealed that, in such cases, multiple paternity of litters can occur (Figure 5.6; Faulkes *et al.*, 1997b). The adaptive significance of multi-male parentage in naked mole-rats is unclear. Given the high genetic similarity of colony members as a result of continous inbreeding (see Chapter 7), it is uncertain what benefit the queen gains by mating with more than one male. One explanation could be that there is a small amount of residual genetic variation within colonies and the queen may maintain some variability in her offspring and reduce the effects of inbreeding by mating with more than one male. This is supported by the fact that some DNA fingerprints show small differences between individuals within a colony (Faulkes *et al.*, 1997b; Chapter 7). It is also possible that the queen may select mates on the basis of genetic differences at the major histocompatibility complex of genes (MHC), as has been demonstrated in mice (Potts *et al.*, 1991). Another explanation could be that as naked mole-rats are so highly inbred, mating with more than one male increases conception rates. As in some other inbred species, the sperm of naked mole-rats are highly polymorphic in shape and size (C.G. Faulkes, unpubl.), and the combination of inbreeding, a high incidence of structurally abnormal sperm, and reduced fertility have been linked in a number of species, e.g. lions (Wildt *et al.*, 1987). Multi-male mating in naked mole-rats also opens up the possibility for sperm competition.

(a)

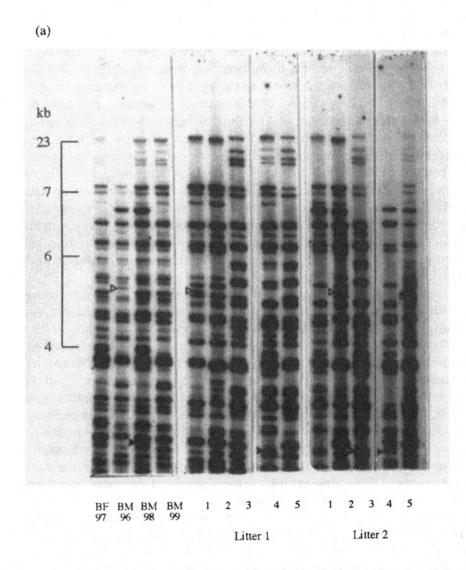

Figure 5.6. A representative DNA fingerprint for captive colony N generated using minisatellite probe 33.6, showing three putative breeding males (96, 98, 99), the breeding queen (97), and two litters of five offspring. In this fingerprint, one band unique to male 96 and one unique to male 98 were detected (open and filled arrow heads, respectively). In both litters offspring could be unambiguously assigned to both of these breeding males (Faulkes *et al.*, 1997b). Paternity testing was possible in this colony due to its abnormally high levels of heterozygosity, produced by crossing genetically distinct northern and southern Kenyan mole-rats (see Chapter 7).

Apart from the breeders, all other males in the colony seem to be oblivious to the presence of a sexually active oestrous female. In the genus *Cryptomys* there are no fights between any of the males for reproductive opportunity, although this occasionally occurs in naked mole-rats (J.U.M. Jarvis, pers. comm.). While there is some degree of pair bond formation between reproductive animals in common mole-rats, it is more developed in Damaraland mole-rats where grooming takes place between the two. In the naked mole-rat there is a strong bond between the breeding queen and the breeding males, with mutual ano-genital sniffing and nuzzling taking place regularly, and the breeding animals often in close body contact while in the confines of the nest chamber (Bennett and Jarvis, 1988a; Bennett, 1989; Jarvis, 1991).

Many of the solitary subterranean rodents appear to be induced ovulators, that is, ovulation is triggered by the act of mating itself rather than occurring spontaneously. The ctenomyid *Ctenomys talarum* is known to be an induced ovulator (Weir, 1974). Altuna and Lessa (1985) have studied the penial morphology of a Uruguayan species of *Ctenomys* and have suggested that the spines on the glans penis may provide cervico-vaginal stimulation for induced ovulation. Indeed, induced ovulation appears to be the general rule in non-gregarious mammals (Zarrow and Clarke, 1968) and would be an adaptive trait in solitary species relying on brief, chance encounters for mating. In the Bathyergidae, it is unknown whether the solitary species are induced or spontaneous ovulators. However, in the social bathyergids, there is evidence to suggest that the common mole-rat may be an induced ovulator (Spinks *et al.*, 1999a). In contrast, the naked mole-rat is known to be a spontaneous ovulator (Faulkes *et al.*, 1990a) and all the available evidence suggests that this may also be the case for the Damaraland mole-rat (A.J. Molteno and N.C. Bennett, unpubl.). These observations show an interesting correlation with the mechanism of reproductive suppression that operates in these species (Chapter 6). In the two eusocial species, physiological blocks to reproduction in the non-breeding females may negate the need for a control of ovulution by the act of mating. In the naked and Damaraland mole-rats, the opportunity for mating is only likely to arise when the individuals are released from the social cues within their colonies that inhibit reproduction, and they become reproductively active. In a sense, release from suppression is acting as an induction. In the common mole-rat, however, there is no physiological suppression of reproduction in females, and induction of ovulation by mating provides the stimulus at the appropriate time.

Ovarian cycle lengths in the Bathyergidae are only known for the naked mole-rat. In this species, the reproductive female has an ovarian cycle of approximately 34 days, with a follicular phase of 6 days and a luteal phase of 28 days (Faulkes *et al.*, 1990a). The comparatively long ovarian cycle is similar to other hystricomorph rodents. For example, in guinea pigs, the corpus luteum of an un-mated non-pregnant female secretes progesterone for 15 to 17 days, and the cycle lengths for other hystricomorphs are comparable, e.g. the cuis, *Galea masteloides* (22 days) the chinchilla, *Chinchilla laniger* (40 days) and the acouchi, *Myoprocta pratti* (30 to 55 days) (Weir and Rowlands, 1974).

5.4 GESTATION AND PARTURITION

Data on gestation length within solitary subterranean rodents are scarce but, where known, show great variability, ranging from 18 to 28 days in the pocket gopher, *Thomomys talpoides* (Schramm, 1961; Andersen, 1978) to 93 to 120 days in the tuco-tuco, *Ctenomys talarum* (Weir, 1974). Members of the Geomyidae, Spalacidae and solitary Bathyergidae have shorter gestation periods than the Ctenomyidae and social Bathyergidae. For example, in the solitary Cape mole-rat gestation is 44 to 48 days and in the Cape dune mole-rat, 52 days, whereas in the social Mashona and common mole-rats, gestation is 56 to 66 days (Bennett, 1989; Bennett *et al.*, 1994a). In *C. amatus* and the giant Zambian mole-rat, *C. mechowi*, gestation can extend to 111 days (Bennett and Aguilar, 1995; Burda, 1989). As with ovarian cycle lengths, a long gestation also seems to be characteristic of hystricomorph rodents (Weir, 1974) and its occurrence in most of the Bathyergidae provides additional support for the placement of the Bathyergidae in the hystricomorph sub-order (Tables 5.1 and 5.2).

Parturition is not commonly observed in subterranean rodents, but in some bathyergids may take place over an extended period, with relatively long intervals between the emergence of pups. Bennett and Jarvis (1988b) reported that a litter of four pups born to a breeding female Damaraland mole-rat took over an hour to be delivered. Bennett (unpublished data) also observed in the giant Zambian mole-rat that three pups were born over a six-hour period, and in one litter of two Mashona mole-rat pups, these were born on consecutive days.

The litter sizes of subterranean rodents are generally small (Bennett *et al.*, 1991; Malizia and Busch, 1991), the exception being the naked mole-rat, which in captivity may have up to 27 pups in a

Table 5.1. A comparison of factors associated with reproduction and parturition for the family Bathyergidae. References as follows: [1]Jarvis, 1969a,b;[2]Van der Horst, 1972; [3]Bennett et al., 1991; [4]Bennett & Jarvis, 1988a; [5]Bennett, 1988, 1989; [6]Bennett & Jarvis, 1988b; [7]Jarvis & Bennett, 1993; [8]Burda, 1989; [9]Bennett et al., 1994a; [10]Bennett and Aguilar,1995; [11]Jarvis, 1991.

Species	No. in burrow	Breeding season: occurence of pregnant and lactating females	Gestation (days)	Max. No. of litters per year	Mean litter size and range	Birth mass (g)	Birth sex ratio M:F (n)	Mean annual recruitment
Bathyergus suillus[1,2,3]	1	Jul–Oct (winter)	c. 52	2	2.4 (1–4)	34	1:1 (6)	5
Bathyergus janetta[3]	1	Aug–Dec	–	2	3.5 (1–7)	15.4	1:3.5 (10)	7
Georychus capensis[4]	1	Aug–Dec	44–48	2	6 (4–10)	5–12	2.5:1 (8)	12
Heliophobius argenteocinereus[1]	1	Apr–Jun	c. 87	–	– (2–4)	7	–	–
Cryptomys h. hottentotus[5]	<14	Oct–Jan	59–66	2	3 (1–6)	8–9	1:1 (6)	6
Cryptomys damarensis[6,7]	<41	All year	78–92	4	3 (1–5)	8–10	1:1 (23)	12
Cryptomys amatus[8]	<25	All year	100	3	2 (1–2)	7.8–8.1	1:1	6
Cryptomys darlingi[9]	<11	All year	56–61	4	1.7 (1–3)	6.9–8.2	1:1	8
Cryptomys mechowi[10]	<8	All year	97–111	3	2 (1–3)	15–21	1:1	6
Heterocephalus glaber[11]	<295	All year	66–74	4	13 (1–27)	1.8	1:1 (>110)	52

Table 5.2. Characteristics of pups at birth, and post-natal development of significant behaviours for the family Bathyergidae. The approximate time of onset of these behaviours is expressed in days after birth (d.a.b.). References as Table 5.1.

Species	Pup length (cm)	Pelage present (d.a.b.)	Eyes open in pups (d.a.b.)	Ear meatus at birth	First eat solids (d.a.b.)	Weaned (d.a.b.)	Begin sparring (d.a.b.)	Disperse (d.a.b.)	Drumming observed (d.a.b.)
Bathyergus suillus[1,2,3]	–	At birth	10	Closed	10	c. 21	12–13	60–65	> 80
Bathyergus janetta[3]	–	4	15	Closed	13	c. 28	11–16	60–65	> 80
Georychus capensis[4]	3–4	7	9	Closed	17	c. 28	35	55–60	50
Cryptomys h. hottentotus[5]	8–9	8	13	Closed	10	c. 28	10–14	Social	–
Cryptomys damarensis[6,7]	8–9	6	18	Closed	6	c. 28	18–25	Social	–
Cryptomys amatus[8]	5–6	9	24	Closed	22	c. 82	–	Social	–
Cryptomys darlingi[9]	–	4	14	Closed	14	c. 45	36	Social	–
Cryptomys mechowi[10]	–	2	6	Closed	20	c. 42	60	Social	–
Heterocephalus glaber[11]	–	No pelage	30	Closed	14	c. 24	21	Social	–

litter (Jarvis, 1991). In the wild, litter sizes estimated from capture of weaned young average around 10 and therefore, at birth, may be slightly higher (Brett, 1991b). This large potential litter size is perhaps only surpassed among mammals by the tenrec, an insectivore, where litters of 32 have been reported (Macdonald, 1984).

Among the other bathyergids, litter sizes and recruitment are somewhat variable, even within genera. When looking at recruitment of young, there is also the issue of sociality to take into account, and the seasonality in breeding. For example, while the naked mole-rat may produce large litters, only one female per colony is breeding and therefore the colony is effectively acting as a single breeding unit. The seasonally breeding bathyergids, including all the solitary species and the social common mole-rat, normally produce one or occasionally two litters annually (Jarvis, 1969a,b; De Graaff, 1981; Bennett and Jarvis, 1988a; Bennett, 1989; Bennett *et al.*, 1991; Malizia and Busch, 1991). In both common and Damaraland mole-rats, the single reproductive female in each colony may produce between one and six pups per litter (Bennett and Jarvis, 1988b; Bennett, 1989), with a mean litter size of three. In contrast, the tropical social Mashona mole-rat, giant Zambian mole-rat and *C. amatus* produce fewer pups, normally one to three (Bennett *et al.*, 1994a; Bennett and Aguilar, 1995; Burda, 1989). Thus, the seasonally breeding common mole-rat produces a maximum of two litters per annum which results in an average annual recruitment of six pups. In contrast, the Damaraland mole-rat, although only producing a mean litter size of three, is an aseasonal breeder with the capacity to produce four litters per annum, which would represent a recruitment of 12 pups. The tropical species of *Cryptomys*, although being aseasonal breeders, because of their small mean litter size essentially have the same annual recruitment as the common mole-rat. As we have mentioned, the naked mole-rat is exceptional within the family in the large size of its litters and the potential recruitment (Jarvis, 1991). This potential to bear unusually large litters may enable the naked mole-rat, under optimal conditions, to capitalise on years of good rainfall or unusually large resource patches that may be encountered. If we consider that a colony's single breeding queen may produce around 50 pups in a year, with an average colony size of 40 to 100 animals, this would equate to an equivalent recruitment of up to around one pup per female (assuming the sex ratio in the colony is equal). In the Damaraland mole-rat, which inhabits a similar environment, an average size colony of 15–18 individuals produces a maximum recruitment of 12 pups, resulting in a comparable figure of 1 to 1.5

pups per female in the colony (Table 5.1).

In marked contrast, in the solitary bathyergid species of the genera *Bathyergus* and *Georychus*, each adult female in the population has the possibility of breeding twice a year, and each may produce between five and 12 pups per year. Among non-bathyergid mole-rats, the tuco-tuco, *C. talarum*, produces between one and eight pups per litter, with a mean of four (Malizia and Busch, 1991). The same is true of *S. ehrenbergi* which has been suggested as a seasonal breeder, the mean litter size being around three to four pups (Nevo, 1961; Shanas *et al.*, 1995). Thus, looking at recruitment in this way, in these solitary species of bathyergid and other genera of subterranean rodents, the annual recruitment per female far exceeds that of the social genera of the Bathyergidae. Females will always try to maximise their genetic contribution to future generations and thus consequently should produce as many young as possible. Interestingly, social mole-rats in which reproductive skew occurs produce comparable numbers of offspring to those of their solitary counterparts. The exception to this rule is the naked mole-rat, in which the sole reproductive female produces more offspring per annum than any other members of the Bathyergidae so far studied.

5.5 SEX RATIOS

In captive naked mole-rat colonies there is a 1:1 sex ratio at birth. However, when all data collected on wild-caught naked mole-rats are pooled, there is a clear bias towards adult males in the colonies ($n = 1,136$, $\chi^2 = 7.78$, $p < 0.01$). But why should the colonies be male-biased? The answer remains unclear but it could mean that more females than males emigrate from the colony. However, this explanation is unlikely since, in most mammals, it is the males that disperse, and evidence suggests that this is the case in naked mole-rats (O'Riain *et al.*, 1996), although females also may disperse (S.H. Braude, pers. comm.). Another possibility is that females engage in tasks that expose them to a higher risk of predation, but there is no supporting evidence of sex-linked division of labour in captive colonies. Perhaps the best explanation is that there are times when there is heightened aggression between the breeding female and other large females in the colony. In the laboratory, if this escalates into full-blown fighting, it often results in the death of one or both of the aggressors and, if this happens often enough over time, it could skew the sex ratio of the colony towards males (Brett, 1991b; Jarvis, 1991;

Lacey and Sherman, 1991).

A somewhat similar pattern of changing sex ratios occurred in a population of Damaraland mole-rats from Namibia. When animals of all sizes were considered, it was found that the population was male biased (592M:461F; χ^2 = 16.3, d.f. = 1, p < 0.00005; Jarvis and Bennett, unpubl. data). Sex ratios were equal in younger animals (body mass <130 g) and only became skewed towards males in those animals weighing more than 131 g. In the Damaraland mole-rat it is unlikely that the deviation in sex ratio from parity is due to the same reason as in naked mole-rats, as an incest taboo exists. Thus, non-reproductive females do not pose a threat to the breeding female in an established colony and inter-female aggression and fighting do not occur. The reason for the male bias in Damaraland mole-rats therefore remains unclear.

The situation in the common mole-rat is even more difficult to understand, because very different sex ratios were found in two populations, one from a mesic and the other from an arid part of their distribution. Thus, there was a bias towards males (510M:228F; χ^2 = 107.6, p < 0.00001) in the population from a mesic area, but a sex ratio near to parity in a population from an arid area (323M:281F; χ^2 = 2.92, p = 0.1; Spinks, 1998).

5.6 PUP DEVELOPMENT AND GROWTH

Among the Bathyergidae, pup development and growth might be expected to be influenced by the social system. In the solitary species aggression increases between the mother and her offspring over time and juveniles disperse after around 60 days (Table 5.2). In contrast, all of the social Bathyergidae absorb the pups into the natal colony. These 'recruits' are readily accepted by the reproductive and non-reproductive colony members, and contribute towards: (i) the direct care of pups, (ii) extending the burrow system, (iii) defending the burrow system, and (iv) replenishing the food store (Chapter 4). The non-reproductive caste in the social mole-rats in essence frees the reproductive animals from activities that carry an increased risk of predation (like foraging in the periphery of the burrow, or kicking soil out of molehills), and allows them to focus on reproduction.

As a consequence of early dispersal, the pups might be expected to be relatively more precocial in the solitary species than the social species. On the other hand, one could also argue that in the social species, where ecological constraints are higher and cooperative behaviour occurs, precocial young would be available for 'work' sooner. However, while there are distinct behavioural differences among the pups of the various species and subspecies, there are no clear trends in pup growth and development to support one or the other of these hypotheses (Bennett *et al.*, 1991; Table 5.2).

The pups of social mole-rats do tend to be more mobile after birth. Pups wander out of the nest for a period of greater than two minutes at an earlier age (day 1 to day 5) than the pups of solitary animals (day 5 to day 9). The pup's agility, or truly coordinated walking movement along the burrow system, develops at much the same time in all the bathyergids. The pups of the Damaraland mole-rat are remarkably precocial and begin to eat solids at a very early age (day 6). Compare this with that of the other bathyergid and subterranean rodents, where this does not occur until day 10 to day 22 (Tables 5.2 and 5.3). Weaning of pups amongst most of the bathyergids occurs at four weeks, with the exception of *C. amatus* from Zambia, where pups are weaned much later at around 82 days (Burda, 1989). Sparring, an interactive behavioural act important in pup development (see Chapter 4), involves wrestling and incisor fencing. It manifests itself early in the solitary Cape and Namaqua dune mole-rats (Table 5.2), but it occurs later, on day 14 to 21, in the more social cryptomids and the naked mole-rat. Surprisingly, sparring occurs only after five weeks in the solitary Cape mole-rat (Table 5.2). In all solitary bathyergid species, sparring intensifies and becomes aggressive with the infliction of wounds if pups are unable to disperse. Hindfoot drumming, used in territorial advertisment, is first exhibited in Cape mole-rat pups around day 50, whereas in the two *Bathyergus* species it occurs around day 80.

Information pertaining to the development in solitary species of other subterranean rodents is scanty (Table 5.3). Again, overall patterns in growth and development are hard to discern. However, rates of development comparable to those of the Cape mole-rat have been found in two unrelated species of comparable mass. Both *Thomomys talpoides* (Andersen, 1978) and *Tachyoryctes splendens* (Jarvis, 1969a) have pups which begin to eat solids at 17 days and open their eyes when 21 to 28 days old. Sparring starts at five weeks and severe fighting amongst pups breaks out at eight weeks in *T. talpoides* and at 11 weeks in *T. splendens*.

In the ctenomid *C. talarum* and the bathyergid *B. suillus* (Cape dune mole-rat), pups possess a pelage at birth while, in contrast, pups of the solitary bathyergids *B. janetta* (Namaqua dune mole-rat) and *G. capensis* (dune mole-rat), the spalacid *S. ehrenbergi* and the

Table 5.3. Morphological and behavioural development of pups in non-bathyergid subterranean rodents, values expressed as days after birth (d.a.b.) where indicated. References as follows [1]Weir, 1974; [2]G.H. Aguilar, unpubl.; [3]Nevo, 1961; [4]Shanas et al., 1995; [5]Gazit et al., 1996; [6]Anderson, 1978; [7]Jarvis 1969a,b and [8]Jarvis, 1973.

Species	Family	Social status	Gestation period (days)	Litter size	Body mass of pups at birth (g)	Pelage present (d.a.b.)	Eyes open in pups (d.a.b.)	Pups leave nest (d.a.b.)	First eat solids (d.a.b.)
Ctenomys talarum[1,2]	Ctenomyidae	Solitary	93–120	5	8.0	1	2–3	7	–
Spalax ehrenbergi[3,4,5]	Spalacidae	Solitary	28–34	1–5	5.0	7–8	Sub-cutaneous eyes	12–14	14–21
Thomomys talpoides[6]	Geomyidae	Solitary	18	–	2.7–3.6	9	26	9	17
Tachyoryctes splendens[7,8]	Rhizomyidae	Solitary	36–41	–	11–18	2–4	21–28	20	15–21

Table 5.4. Growth characteristics for eight species of the family Bathyergidae. With the exception of the first two columns, the other parameters are derived after fitting the Gompertz model for growth to the empirical data; the asymptotic weight corresponds to where the modelled growth curve flattens off, the mean maximum growth rate is calculated by multiplying the growth constant by the A function of the asymptote, and the inflection time is extrapolated from the point on the growth curve where the the curvature alters. References as follows [1]Bennett et al., 1991; [2]Bennett and Aguilar, 1995; [3]Bennett et al.,1994a; [4]Jarvis, 1991; [5]Brett, 1991a and [6]O'Riain, 1996.

Species	Mean body mass at birth (g)	Mean adult body mass (g)	Mean maximum growth rate (g day⁻¹)	Projected time to attain mean adult mass (days)	Asymptotic weight (A)	Inflection time (days)	Mean growth weight constant (K)	Maximum growth rate (g day⁻¹)
Bathyergus suillus[1]	30.0	780	3.340	227	217.5	22.3	0.04	3.300
Bathyergus janetta[1]	15.0	390	1.638	223	90.8	14.8	0.05	1.680
Georychus capensis[1]	6.0	180	1.227	143	74.6	16.9	0.04	1.220
Cryptomys h. hottentotus[1]	8.0	67	0.229	268	42.0	12.6	0.015	0.220
Cryptomys mechowi[2]	18.0	272	0.847	299	90.9	26.7	0.025	0.850
Cryptomys darlingi[3]	6.2	65	0.272	216	92.6	94.1	0.008	0.272
Cryptomys damarensis[4]	9.0	100	0.233	436	42.5	15.6	0.015	0.230
Heterocephalus glaber[5,6]	1.8	33	0.207	Variable	Variable	Variable	0.017	Variable

geomyid *T. talpoides* are naked and helpless at birth. The pups of *C. talarum* and *S. ehrenbergi* are generally larger (8 g) than those of *T. talpoides* (3 g). In the latter two species, the pelage arises between three and 14 days. The eyes of all species are closed at birth and they remain in the nest for around 9–14 days. The pups of *Spalax* and *Thomomys* begin eating around 14–17 days, an age comparable to that of the Cape mole-rat (Table 5.3).

There is a general trend for the pups of the Bathyergidae to be born in an altricial state and develop slowly, in common with other subterranean rodents (Bennett *et al.*, 1991). The nidicolous condition of the pups in the family Bathyergidae, as well as the other families of subterranean rodents, may in part be related to the thermally stable and relatively secure environment of the burrow (Bennett *et al.*, 1991).

Both the mean growth rate and the maximum growth rate (Table 5.4) are higher in the solitary species, but in the case of the Cape dune mole-rat and Namaqua dune mole-rat it could be argued that this is because they have a larger adult body size. However, this is not the case with the Cape mole-rat, which shows faster growth but is smaller than the social giant Zambian mole-rat. There is no apparent trade off between the production of precocial offspring and the size of the litter amongst mole-rats. The young of most species are born relatively altricial and yet litter size can be variable. Interestingly, an intra-specific study on litter size and pup body mass has revealed that in litters which are small, the pups tend to be much larger and stronger than those of larger litters produced by the same mother (R. Cooney, pers. comm.). Among birds, a similar finding has been made in the great tit, *Parus major*, where larger broods of great tits weigh less at fledging because the parents cannot feed them as efficiently (Lack, 1966). The general take home message here is that heavier chicks survive better. Hence, in the great tit, the optimal clutch size is the one which maximises the number of surviving young per brood and so maximises lifetime breeding success rather than that which maximises success per breeding attempt (Krebs and Davies, 1991).

The differences in adult body size of the different bathyergid species are reflected in differences in growth characteristics (Table 5.4). The various parameters displayed in Table 5.4 were derived from Gompertz sigmoidal growth models fitted to empirical growth curves, enabling intra- and inter-specific comparisons to be made (Zullinger *et al.*, 1984). The solitary mole-rats (the Cape and Namaqua dune mole-rats) for the first 70 to 80 days of post-natal growth have similar mean maximum growth rates of 1.2 and 1.7 g per day, respectively. For the Cape dune mole-rat, the largest of the Bathyergidae, the mean maximum growth rate is 3.3 g per day (Bennett *et al.*, 1991; Table 5.4). The mean rates of growth in these solitary bathyergids are comparable to the values obtained for other comparatively sized solitary subterranean and fossorial rodents. For example, *Thomomys ruandae* has a mean growth rate of 1.3 g per day (Rahm, 1969), *T. talpoides* 2.2 g per day (Andersen, 1978) and *Tatera brantsii* 1.7 g per day (Meester and Hallett, 1970). Similarly, the prairie dog, *Cynomys ludovicianus*, has a mean growth rate of 4.0 g per day (Anthony and Foreman, 1951) which is comparable to that of the similar sized Cape dune mole-rat.

The social common and Damaraland mole-rats have similar mean maximum rates of growth of 0.23 g per day. Burda (1989), using linear regression models, calculated the rate of growth in *Cryptomys amatus* from Zambia to be 0.21 g per day in females and 0.3 g per day in males for the first 80 days after birth, which is comparable to the findings found in the other species of *Cryptomys*. The time taken to attain adult body mass is comparatively long in the Bathyergidae (Table 5.4), although there is a general trend for the solitary species to attain adult mass at an earlier age than those of the social species. Again, this may reflect their different life history and reproductive strategies, and may also be partly due to the lower metabolic rates recorded in the genus *Cryptomys* (Table 5.4; McNab, 1966; Bennett *et al.*,1992, 1993b, 1994b). The possession of a secure nest underground may have promoted slow development of pups in the Bathyergidae, as it has with other altricial mammals known to nest exclusively below ground, in caves or in tree hollows (Case, 1978).

As with many other aspects of their biology, the growth characteristics of naked mole-rats also stand apart from the other bathyergid mole-rats, and their complex social system seems to be reflected in patterns of growth that are also complex, and show considerable variation. As we have mentioned in Chapter 4, behavioural role and dominance status within naked mole-rat colonies are closely linked to both body size and age (Jarvis *et al.*, 1991; Lacey and Sherman, 1991; Schieffelin and Sherman, 1995; O'Riain, 1996; Clarke and Faulkes, 1997). However, with the exception of very young animals and individuals in the first two litters of a nascent colony, age and body size do not appear to co-vary in a simple way in all individuals, and there are differences both within and between colonies. Long-term studies of growth in captivity have shown that, although the mean growth rate constant remains the same for all naked mole-rats, individuals show differences in other growth characteristics like asymptotic weight, inflection time and the projected time to attain adult mass (Table 5.4; Jarvis *et al.*, 1991; O'Riain, 1996). In addition, naked mole-rats show the slowest mean maximum growth rate during the first 80 days of any of the Bathyergidae so far studied (O'Riain, 1996). This slow growth may be related to the low metabolic rate and poikilothermic thermoregulatory traits of naked mole-rats (Withers and Jarvis, 1980; Buffenstein and Yahav, 1991b). The plasticity in growth can be clearly seen if a new colony is founded in captivity from a pair, where the animals born in the first few litters attain a significantly greater asymptotic body mass than those in later litters. Also, the pups within the first litter usually show a greater mean maximum growth rate, even with *ad lib* access to food (O'Riain, 1996). This plasticity in growth may serve to quickly recruit into the colony individuals with a full range of body sizes, and thus provide individuals with a full range of behavioural roles. We have already seen in Chapter 4 that 'work' or 'colony maintenance' tasks are negatively correlated with body mass, while 'defence' related activities increase with body mass (body size polyethism). Slowing the growth of successive litters would also function to maintain these behavioural divisions of labour, and prevent the colony from becoming full of large individuals. In turn, this would also have a benefit in terms of a lower energetic cost, and the ergonomics of the colony can remain optimal (O'Riain, 1996). Alternatively, it could be argued that older individuals (i.e. those of the first and second litters) suppress the growth of individuals in younger litters and thereby increase their chances of becoming the dominant reproductive. This appears to be true in the case of the naked mole-rat (Clarke and Faulkes, 1997; 1998). Further evidence for the link between body mass and energetic constraints can be seen in size differences between southern and northern Kenyan wild-caught animals. While in the colonies around Tsavo and Mtito Andei in southern Kenya, mean body mass is 33 g (Brett, 1991a), further north in Kenya at Lerata, where food availability is lower, mean body mass drops to 23 g (Jarvis, 1978; Jarvis *et al.*, 1991). The flexibility in growth patterns and lack of morphological specialisation among non-breeders means that naked mole-rats can show considerable behavioural plasticity in response to changing social and environmental conditions. This ability to optimise the behavioural and morphological profile of the colony may well be adaptive as tasks can be matched to body size, and colony efficiency, and therefore inclusive fitness of individuals, can be maximised (O'Riain, 1996).

Interestingly, while body size may remain constant in some individuals for prolonged periods of time, profound and quite sudden body mass changes, or 'growth spurts', may occur in response to changes in the social environment of the colony, particularly following death or removal of breeding animals (O'Riain, 1996; Lacey and Sherman, 1997; Clarke and Faulkes, 1997, 1998). In one captive colony where the breeding queen was killed during a 'coup' by a non-breeder who became reproductively active, challenged and successfully deposed the queen, the usurper had dramatically increased her body mass prior to the event. In the space of about 70 days this adult female increased from 35 g to 55 g, whereas in the 120 days before the onset of this growth spurt, she had only gained 4 g in mass (Faulkes, 1990). Following the attainment of breeding status, the queen's vertebrae gradually elongate to produce the characteristic body size and shape of a breeding female (Jarvis *et al.*, 1991; O'Riain, 1996). It is the older litter members that show greater growth responses to the death of a breeding male, queen, or older siblings, again emphasising the relationship between age, body size, dominance and reproductive success. We know that behaviour is correlated with body size, and to some extent age. Also, dominance is significantly correlated with body mass and the attainment of reproductive status (Clarke and Faulkes, 1997, 1998) and it seems likely that dominance, growth, behavioural role and reproductive success may all be mechanistically linked.

The ability of 'adult' naked mole-rats to recommence growth is extremely unusual in mammals, where, after puberty, further growth ceases when the epiphyses, the plates at the end of bones where they increase in length, fuse with the rest of the bone, the diaphyses. This is, however, consistent with the idea that puberty is effectively blocked in nearly all colony members, resulting in the reproductive suppression of all but the breeders, a subject that we will discuss fully in the next chapter. The growth spurts that are seen in some individuals, for example, when a breeder dies, may then be analogous to the pubertal growth spurt in other mammals, as these individuals also become reproductively active. The growth characteristics of naked mole-rats are another example of phenotypic plasticity (see also Chapter 4) set against a background of genotypic similarity (see Chapter 7), and further emphasise the relative contribution of 'nurture' (environmental upbringing) versus 'nature' (the genetic contribution to phenotype).

5.7 DISPERSAL

Patterns of dispersal among the Bathyergidae are critically important factors to consider if we are to understand the origin and maintenance of their social systems. Constraints on dispersal influence the degree of reproductive skew, and the subsequent group sizes and degree of sociality. The inter-relationships between these factors, together with the patterns of genetic relatedness within and between colonies and ecological factors such as aridity, will be discussed fully in Chapters 6, 7 and 8.

The solitary species in the Bathyergidae are forced to leave the maternal burrow and establish their own territory at an early age. In the Cape mole-rat, the juveniles disperse either by extending the maternal burrow system or by moving short distances above the ground (Jarvis and Bennett, 1991). Limited data for the Cape dune mole-rat indicate that these mole-rats disperse at a similar age and in much the same fashion as the Cape mole-rat (E. McDaid, unpubl.). Thus, it is quite easy to see that populations will exhibit low vagility and consist of closely related individuals. Among solitary bathyergids, mothers become very intolerant towards their juveniles and it is believed that it is they who ultimately 'decide' when it is time for the pups to disperse. At present, little else is known about the dispersal patterns of the solitary bathyergids or other subterranean rodents, although an interesting observation has been made in the ctenomyid *C. talarum*. In this species the sex ratio at birth and among the the younger animals is 1:1. However, in the adult population there is a skew towards females. It is suggested that during dispersion of the sub-adults, looking to establish their own territories, females disperse prior to males, the consequence of this being that the females would occupy territories more immediate to the maternal system, whereas the males are forced to move long distances above ground in search of territories and, as a result, are at an increased risk of predation (Malizia and Busch, 1991).

While juveniles of the solitary species all disperse as soon as they are mature enough, in the social species of the Bathyergidae they remain in the natal burrow until the opportunities for dispersal arise. This may be relatively soon in species like the common mole-rat, where dispersion may occur every year, or not at all for some naked mole-rats. These windows of opportunity for the mole-rats to disperse from the natal colony, and establish nascent colonies, seem to arise during periods of good rainfall when the soil can be more easily worked. Field studies on Damaraland mole-rats at a site in

Namibia have shown that after a period of rainfall many new colonies were formed. However, the future success of these new groups appeared to be critically dependent on the size of the newly founded colonies. Eight out of 16 colonies that had only two to four adults failed within two years, compared with only one out of eight new colonies that had over eight individuals and an established workforce (Jarvis *et al.*, 1994).

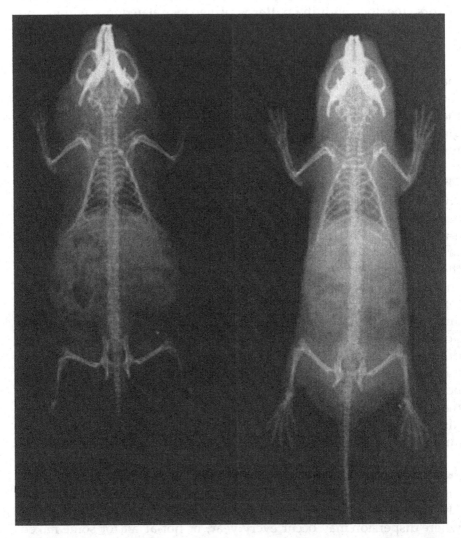

Figure 5.7. X-ray photographs of a naked mole-rat disperser (left) compared with a littermate of similar body size and age (right). Note the significantly greater girth-to-body-length ratio and increased fat in neck region in the disperser. Reprinted with permission from *Nature* (O'Riain *et al.*, 1996) © (1996) Macmillan Magazines Ltd.

In naked mole-rats, dispersion and outbreeding were thought to be extremely rare events in the wild, because the constraints on successful dispersal are so high, although it has been observed on occasion (Sherman *et al.*, 1992). More recently, a specific disperser phenotype, or 'disperser morph', was discovered in some captive colonies of naked mole-rats at the University of Cape Town (O'Riain *et al.*, 1996). Jarvis and O'Riain noticed over a number of years that it was certain individuals that had a propensity to escape from their natal colony and invade adjacent foreign burrow systems. When these individuals, which were nearly always male, were examined more closely, they were found to be morphologically, behaviourally and physiologically distinct. Although 'non-breeding males' (see Chapter 6), these disperser morphs, like breeding males, had elevated plasma concentrations of pituitary luteinising hormone. They were also found to have a significantly greater percentage of body fat compared with non-dispersers from the same age cohort, perhaps to act as a reserve during dispersal (Figure 5.7). Finally, these individuals do not show the usual xenophobic responses of naked mole-rats, and instead solicit matings with non-colony members (O'Riain *et al.*, 1996). An interesting twist to the story was the suggestion by Dawkins in his later edition of *The Selfish Gene* (1989), surprised at the apparent lack of dispersal in naked mole-rats, that a disperser would one day be discovered. He speculated, 'My hunch is that one day we shall discover a dispersal phase which has hitherto, for some reason, been overlooked. It is too much to hope that the dispersing individuals will literally sprout wings!' (Dawkins, 1989). Joking aside, Dawkins was correct in his speculation, and a few years later the morphologically distinct disperser was found (O'Riain *et al.*, 1996). The factors that trigger these rare cases of naked mole-rat dispersal, and the frequency of such events in the wild, are at present unclear. The occurrence of multiple breeding males and male dispersers in naked mole-rats means that males have a greater opportunity for individual reproduction compared with females, and this may explain the lower levels of aggression seen during succession of breeding males (Clarke and Faulkes, 1998).

5.8 LONGEVITY OF BREEDING AND LIFETIME REPRODUCTIVE SUCCESS

Many solitary subterranean rodents have relatively short periods during which they can breed when compared with solitary and social bathyergids. In the Geomyidae, females tend to outlive males. Thus while male *Pappogeomys castanops* and *T. bottae* can live for 31 weeks, the females can live for 56 weeks in the case of *P. castanops* and 55 weeks in *T. bottae* (Howard and Childs, 1959; Smolen *et al.*, 1980). Likewise in the Ctenomyidae, the lifespan of the breeding *C. talarum* is less than 22 months (Busch *et al.*, 1989; Malizia and Busch, 1997). There are no longevity records for the spalacid, *S. ehrenbergi* (Gazit *et al.*, 1996). A more extensive database is currently available for the Bathyergidae. In social bathyergids, residency of reproductives is exceptionally long for small mammals. Records for wild naked mole-rats have shown the same male and female sole breeders in the colony for 10 years (S.T. Braude, pers. comm.), while records for captive breeding male and female tenure exceed 15 years (J.U.M. Jarvis, pers. comm.). Studies on the Damaraland mole-rat have shown the reproductive pair to have resided and bred in the natal burrow for eight years and non-breeding animals as long as four years (N.C. Bennett and J.U.M. Jarvis, unpubl.). Similarly, in the common mole-rat the reproductive pair have been shown to reside in the colony for at least four years (N.C. Bennett, A.C. Spinks and C.M. Rosenthal, unpubl. data). Data on solitary mole-rats are absent. However, a female Cape mole-rat produced litters over four consecutive years in captivity (N.C. Bennett, unpubl.).

In general there is a major paucity of data on lifetime reproductive success in all families of subterranean rodents. Within the Bathyergidae, long-term studies on naked mole-rats have revealed that less than 0.1% of non-breeding mole-rats eventually attain breeding status and can produce a maximum of 48 pups per annum (Jarvis *et al.*, 1994), whereas field studies on the Damaraland mole-rat have shown that ecological constraints limit opportunities for dispersal and approximately 10% of all non-reproductives attain reproductive status with the potential of producing 12 pups per annum (Jarvis and Bennett, 1993).

Chapter 6

Social suppression of reproduction in African mole-rats

6.1 AN OVERVIEW OF REPRODUCTIVE SUPPRESSION IN MOLE-RATS AND OTHER MAMMALS

In the previous chapter we examined the life history patterns and reproductive strategies of the Bathyergidae, and introduced the phenomenon of the reproductive division of labour and socially-induced suppression of fertility seen in the social bathyergids. In many rodents, modulation and suppression of reproductive function, mediated by urinary semiochemicals, is an adaptive response to environmental factors such as increased population density (Brown and MacDonald, 1984). Reproductive suppression is also common in hierarchial groups of competitively and cooperatively breeding mammals, where dominant individuals may inhibit the reproduction of subordinates via behavioural, and other, interactions (Abbott, 1987; Abbott et al., 1988). Social competition and reproductive suppression in subordinate individuals play a major role in determining individual reproductive success in such mammalian species. Examples of social suppression of reproduction reach an extreme in singular cooperatively breeding species, i.e. societies in which auxiliaries do not reproduce while helping (Brown, 1987); these include South American primates (Abbott, 1984, 1987; Tardif, 1997), canids (Asa, 1997), and mongooses and meerkats (Rasa, 1973; Creel and Waser 1997; Doolan and Macdonald, 1997). In these species

157

reproduction is normally restricted to a single female and one or two males. Because of this reproductive division within social groups, these species are often referred to as having a 'high reproductive skew', as opposed to low skew societies where most individuals have an equal chance of reproduction. Theories of optimal reproductive skew aim to explain, in a unified way, the observed range of skew levels seen in different species of cooperative breeders, taking into account the genetic relationships within groups and factors such as environmental constraints (Vehrencamp, 1983; Keller and Reeve, 1994; Sherman *et al.*, 1995). We will consider aspects of reproductive skew theory in more detail in Section 6.7.

The mechanism by which reproduction is suppressed amongst female subordinate mammals may vary. In extreme instances reproduction can be completely suppressed in that ovulation is blocked. Such extreme infertility in subordinate females occurs in high skew cooperatively breeding societies, when the reproductive female requires help with raising the offspring, hence maximising their chance of survival. Because of this, constraints on individual reproduction by non-breeders are high (Wasser and Barash, 1983; Sherman *et al.*, 1995).

In relinquishing reproduction, subordinate males and females forego direct fitness benefits. However, social groups of cooperative breeders are often composed mainly of close relatives, for example, parents and overlapping generations of offspring (all first degree relatives), as in most colonies of naked mole-rats, *Heterocephalus glaber*. Therefore, the non-breeding animals gain indirectly by accruing inclusive fitness benefits, whereby their genes which are shared by their relatives are carried into the next generation by the latter breeding, rather than directly by their own reproduction. In the aforementioned case of the naked mole-rats, non-reproductives have the same degree of relatedness (the average number of genes shared) with the offspring of the reproductive female as they would their own offspring. This is normally around 0.5 for first degree relatives, due to the approximately equal contribution of genetic material from eggs and sperm and the patterns of inheritance described by Mendel's Laws. This figure can reach a maximum of 1.0, or 100% of genes shared, in monozygotic twins. Due to the continued cycles of inbreeding that naturally occur in naked mole-rat colonies (see Chapter 4), the average relatedness is elevated from 0.5 to around 0.8, the highest value so far recorded for a mammal in the wild (Reeve *et al.*, 1990). Despite the high levels of relatedness seen in naked mole-rats and the obvious inclusive fitness benefits for the non-breeders, the inbreeding may be a derived trait peculiar to this species, arising from the high costs and constraints on dispersal. All the other species of bathyergids that have been studied have a strong inbreeding avoidance mechanism yet retain high levels of reproductive skew. Thus, in most social mammals, non-reproductive subordinate helpers benefit from delaying their own reproduction until conditions are optimal for them to either disperse to found a new colony, or succeed as a reproductive in their own group, or perhaps attain this position in another social group as may occur in common marmoset monkeys, *Callithrix jacchus* (Abbott, 1987), and disperser morphs in naked mole-rats (O'Riain *et al.*, 1996). The relationships between skew, ecological factors and relatedness will be considered in detail in Chapter 7.

In high skew cooperatively breeding primates (Abbott, 1987) and herpestids, like the dwarf mongoose (Creel and Waser, 1997), reproductive suppression may not be complete and female subordinates may occasionally produce offspring, though at a lower frequency than the dominant breeding female. Subordinate males may also sire litters. For example, in a study where parentage of dwarf mongooses was determined by DNA fingerprinting, 15% had subordinate mothers, and 25% had subordinate fathers (Keane *et al.*, 1994). In mole-rats, dual-queened colonies have been reported in colonies of captive and wild naked mole-rats (Jarvis, 1991; Sherman *et al.*, 1992) and in wild common mole-rat colonies, *Cryptomys h. hottentotus* (A.C. Spinks, N.C. Bennett and C. Rosenthal, unpubl.). In both these species of mole-rat, however, this is the exception rather than the rule.

Amongst the Bathyergidae, the restriction of reproduction to a single female and one or two males in the colony was first recorded in the naked mole-rat by Jarvis (1981). This reproductive suppression is long term and captive naked mole-rat colonies have retained the original breeding males and females for more than 15 years (Jarvis, 1991). Braude (pers. comm., 1997), carrying out field work in Kenya, has recaptured the same wild-living breeding male and female for a decade. More recently, restricted reproduction has been found to occur in a number of species of *Cryptomys*, including the Damaraland mole-rat, *C. damarensis* (Bennett and Jarvis, 1988b; Bennett, 1994; Bennett *et al.*, 1994c), the common mole-rat, *C. h. hottentotus* (Bennett, 1989), the Mashona mole-rat, *C. darlingi* (Bennett *et al.*, 1994a), *C. amatus* from Lusaka, Zambia (C.G. Faulkes, J.U.M. Jarvis and N.C. Bennett, unpubl.; Burda, 1990) and the giant Zambian mole-rat, *C. mechowi* (Burda and Kawalika, 1993; Bennett and Aguilar, 1995). As with the naked mole-rat, reproduction is restricted to a single female, but there is usually only one breeding male. In the Damaraland mole-rat, colonies are usually founded from a pair of animals originating from different colonies, or occasionally from an adult female and a small group of sub-adults and adults. Out of 17 new colonies founded at a field study site in Namibia, 14 were begun from a pair of adult individuals (Jarvis and Bennett, 1993). Preliminary studies have also revealed that the founding pairs are genetically divergent, having a different maternal lineage (Faulkes *et al.*, 1997a; Chapter 7).

One fascinating aspect of these differences in mating patterns and of this restricted breeding in mole-rats is that the degree and mechanism of reproductive suppression differs in various species of mole-rat. We will now consider the mechanisms and possible explanations for this continuum of socially-induced fertility in detail.

A number of studies have begun to tease apart the physiological nature of reproductive suppression in both male and female mole-rats in a number of bathyergid species. Anatomical and histological investigation of the reproductive tract has been supported by endocrinological studies that have looked at a number of reproductive hormones, using both one-off 'snapshots' of hormonal levels in different individuals, and longitudinal stuies looking at changing levels across time. Some of the hormonal parameters that have been measured are summarised in Figure 6.1, which shows a highly simplified diagram of the hypothalamic–gonadal axis in a mammal. Gonadal function can be compared in breeders and non-breeders by measuring testosterone in males and oestrogens and progesterone in females, the latter normally being released after ovulation. Higher up the hypothalamic–pituitary–gonadal axis, pituitary function can be compared by measuring the gonadotrophic hormone, luteinising hormone (LH), and by comparing the responses of individuals to the administration of a standard dose of gonadotrophin releasing hormone (GnRH). The natural response to GnRH is the release of LH from the pituitary. This is a useful technique for looking at differences in responsiveness to exogenous GnRH, which arise from differences in the secretion and dynamics of action of endogenous GnRH.

Naked mole-rats may be unique amongst singular cooperative breeders, including the other Bathyergidae, in that there are physiological changes that lead to a block to reproduction among non-breeding males as well as females. Naked mole-rats are perhaps also unique among mammals, in that they continuously inbreed, whereas all other mole-rats appear to have inbreeding avoidance

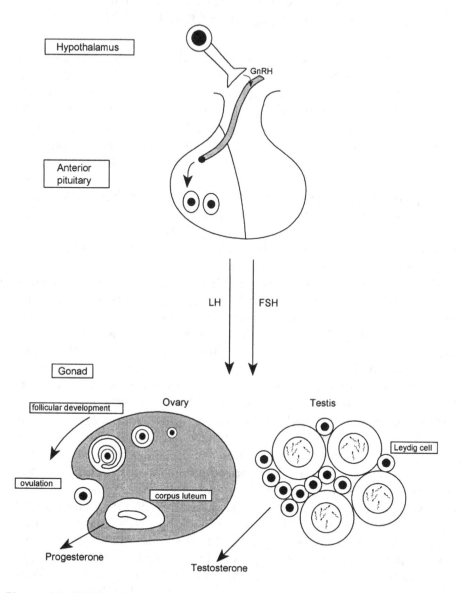

Figure 6.1. Highly schematic diagram illustrating the main components of the hypothalamic–pituitary–gonadal axis that were investigated in breeding and non-breeding mole-rats.

mechanisms, and are obligate outbreeders. Most of the detailed physiological studies have been done on naked and Damaraland mole-rats and these will be considered first before differences between them and the other social cryptomids are discussed below.

6.2 SUPPRESSION OF REPRODUCTION IN MALE NAKED AND DAMARALAND MOLE-RATS

In both captive and wild colonies of the naked mole-rat there are differences in the reproductive tracts of breeding and non-breeding males. The mean testis mass, relative to body mass, of the reproductively active male is significantly larger than that of the non-breeding males (Faulkes *et al.*, 1994). Histological examination reveals that non-breeding males have fewer testosterone secreting Leydig cells (Faulkes, 1990). While there is evidence of spermatogenesis in all male naked mole-rats (Jarvis, 1991), closer investigation has shown that non-breeding males have significantly lower numbers of sperm in their reproductive tract compared with breeders (1.8 x 10^6 versus 8.6 x 10^6 sperm counted in one half of the reproductive tract, respectively). It has also been found that in most non-breeders, these sperm are non-motile (Faulkes *et al.*, 1994).

There are also marked hormonal differences between the breeding and non-breeding males. Endocrine studies have shown that non-breeders have significantly lower urinary testosterone concentrations, low or undetectable concentrations of basal LH and a reduced pituitary response to exogenous administration of GnRH, when compared with breeding males (Faulkes and Abbott, 1991; Faulkes *et al.*, 1991). However, there is considerable variation within a colony. Recent investigations into the basal and post GnRH responses of entire colonies of naked mole-rats, at different phases of the reproductive cycle of the breeding female, show that, at times, some presumed non-reproductive males have concentrations of LH equalling or exceeding the reproductive males. These GnRH responsive males are often the older and larger colony members (Van der Westhuizen, 1997). Additionally, O'Riain *et al.* (1996) found that the disperser morph in naked mole-rat colonies (see Chapter 5) had significantly higher levels of plasma LH than the other non-breeders. The suppressed non-reproductive males will rapidly undergo reproductive activation on removal from their natal colony. When removed, and either housed singly or paired with a female, concentrations of plasma LH and urinary testosterone increase significantly, with urinary testosterone reaching levels comparable to those of breeding males after approximately five days (Faulkes and Abbott, 1991).

Apart from the clear physiological differences between breeding and non-breeding male naked mole-rats, there also appears to be control exerted over the reproductive physiology of breeding males.

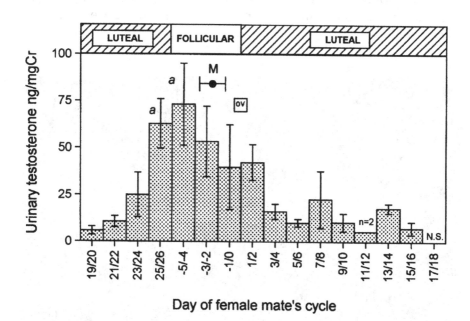

Figure 6.2. Concentrations of urinary testosterone in 68 samples collected from five male naked mole-rats, housed in male–female pairs, over 12 female ovarian cycles. The data are shown as the mean ± S.E.M., plotted relative to the ovarian cycle of the female mate, in two-day intervals, assuming a mean total cycle length of 34 days (Faulkes *et al.*, 1990a). Six observations of mating (M: mean ± S.E.M. day of mating) were coincident with the follicular phase of the cycle just prior to the presumed day of ovulation (ov). N.S.: no samples were collected from this period. *a* = *p* < 0.05 versus days 15–16, 19–20 and 21–22, based on Duncan's multiple-range test following one-way ANOVA for repeated measures (reproduced from Faulkes and Abbott, 1991).

This phenomenon can be seen both in a colony situation and in male–female pairs. In the latter case, when urinary testosterone values of reproductively active males are plotted relative to the ovarian cycle of the female mate, testosterone concentrations in the males peak during the follicular phase of the cycle, just before oestrus and ovulation (Figure 6.2). This reproductive control over the breeding male may help to keep down levels of aggression associated with elevated testosterone, or to reduce the potentially harmful effects of high testosterone titres. The latter may lead to immune suppression, and seasonally-induced high testosterone levels in free-living male hopping mice, *Notomys alexis*, have been implicated in the annual demise of these males following mating (Breed, 1976). Even so, breeding males eventually lose body condition and take on an emaciated appearance, and Jarvis (1991) has shown a consistent loss of 16 to 34% of body mass in these males, in the space of one to seven years.

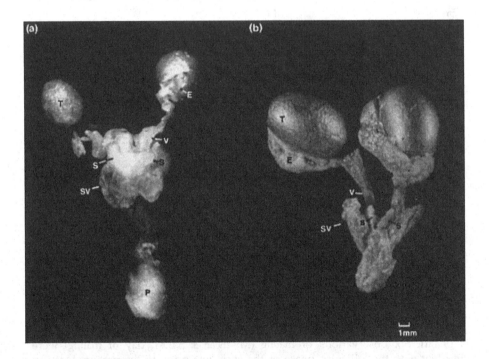

Figure 6.3. Reproductive tract and testis of breeding males in (a) the naked mole-rat, and (b) the Damaraland mole-rat. T: testis; E: epididymis; V: vas deferens; S: sperm storage sac; SV: seminal vesicle; P: penis.

There are clear species differences in the mechanisms of reproductive suppression in non-breeding males of the naked mole-rat and the Damaraland mole-rat, which may be linked to the latter having strong incest avoidance. In the Damaraland mole-rat there is no apparent suppression of sperm production or motility in non-breeding males (Faulkes *et al.*, 1994). The presence of motile sperm in the reproductive tract of all male Damaraland mole-rats also supports endocrine studies that found no differences in concentrations of urinary and plasma testosterone (Bennett, 1994), plasma LH or LH responses to exogenous GnRH between breeding and non-breeding males (Bennett *et al.*, 1993b). Breeding males do, however, have a greater testis mass relative to their body mass, compared to non-breeding males. This gives rise to the characteristic bulging testes (which are not scrotal but sit in inguinal pockets) that distinguish live breeding males from non-breeders (Jarvis and Bennett, 1993). This larger testis, however, does not appear to result in an increased number of spermatozoa in breeders ($0.13 \pm 0.06 \times 10^6$) versus non-breeders ($0.29 \pm 0.14 \times 10^6$) (Faulkes *et al.*, 1994). Males in both Damaraland and naked mole-rats possess characteristic sperm storage sacs (Faulkes *et al.*, 1994; Figure 6.3). In the naked mole-rat these structures are found to enlarge after reproductive activation occurs in non-breeding males removed from their colonies. In reproductively active male naked mole-rats, these structures are packed with sperm. The Damaraland mole-rat therefore conforms to the more usual pattern of a socially-suppressed male mammal, in that there is no obvious physiological suppression, and the failure to breed in non-breeding males apparently results from an exclusion from mating (Bennett *et al.*, 1993b; Bennett, 1994). This may arise simply from an inbreeding avoidance mechanism, because in a typical colony of Damaraland mole-rats, all the non-breeding males will be the offspring of the breeding female (Jarvis and Bennett, 1993; Jarvis *et al.*, 1994; Faulkes *et al.*, 1997a). Conversely, in naked mole-rats, in order to maintain the reproductive division of labour among males, a physiological mechanism seems to be required as they lack the inbreeding avoidance component of Damaraland mole-rats.

6.3 SUPPRESSION OF REPRODUCTION IN FEMALE NAKED AND DAMARALAND MOLE-RATS

In female naked mole-rats, anatomical and histological studies have shown that most non-breeders in the colony have ovaries with a pre-pubescent appearance (Kayanja and Jarvis, 1971; Faulkes, 1990; Faulkes *et al.*, 1990c). In these females, the cortex of the ovary is packed with primordial follicles, but few developing follicles are present and it is rare to find tertiary follicles. Most non-breeding female naked mole-rats are also characterised by low or non-detectable concentrations of urinary progesterone and oestrogen (Faulkes *et al.*, 1991; Westlin *et al.*, 1994), confirming that in both captive and wild colonies of the naked mole-rat, the block to reproduction in non-breeding females results from an inhibition of ovulation. The block to ovulation in these suppressed females is apparently due to an inadequate secretion of LH from the anterior pituitary gland: plasma LH concentrations are significantly lower in non-breeders when compared with the breeding female (Faulkes *et al.*, 1990c). The other pituitary gonadotrophin, follicle stimulating hormone (FSH), may also be inhibited, but levels of this hormone have not been measured in mole-rats. In turn, this reduction in plasma LH may result from a disruption in the normal patterns of secretion of GnRH from the hypothalamus of the brain. Responses of the pituitaries of non-breeding female naked mole-rats to stimulation by administration of exogenous GnRH suggest a lack of sensitivity to GnRH (Faulkes *et al.*, 1990c, 1991; Figure 6.4a). This could be due to a reduction in pituitary GnRH receptors or an alteration in the post receptor metabolic events in non-breeders. In rats, it has been shown that changes in the pituitary responsiveness to exogenous GnRH during the oestrous cycle are paralleled by changes in the pituitary GnRH receptor concentrations (Smith, 1984). It is possible that in non-breeding female naked mole-rats the anterior pituitary has reduced concentrations of GnRH receptors because endogenous release of GnRH is disrupted, as the ability of GnRH to autoregulate its receptors is well known (Sandow, 1983).

By experimentally priming the pituitary of non-breeding females with as little as four consecutive hourly doses of GnRH, it is possible to reverse the lack of responsiveness, presumably by up-regulation of the GnRH receptors on the pituitary gonadotroph cells (Faulkes *et al.*, 1990c). The reversible nature of the lack of responsiveness reveals the rapidity with which non-breeding female naked mole-rats can be released from their anovulatory condition. Fertility is normally completely inhibited in non-breeding female naked mole-rats while they remain in the confines of the colony. However, if removed from the socially suppressing environment, by separating females from their parent colony and either housing singly or pairing with a male, ovarian cyclicity may start as early as around eight days post removal (Figure 6.5). This is surprisingly fast considering that non-breeding females may have been suppressed for many years, and may be akin to going through puberty in the space of a few days!

The activation of ovarian cyclicity in singly housed females means that the presence of a male, or mating, is not required for ovulation, and thus, naked mole-rats are spontaneous ovulators (Faulkes *et al.*, 1990a,c; Figure 6.5b). This rapid onset of reproductive activity in non-breeding females can also occur within a colony context, after the death of the breeding female. Jarvis (1991) documents an instance when the successor to a queen mated within eight days of her death and had her first litter 74 days later (the gestation length of naked mole-rats).

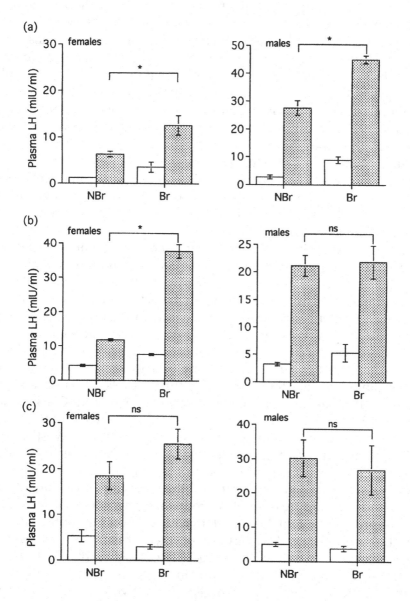

Figure 6.4. The socially-induced infertility continuum in the three mole-rats, (a) the naked mole-rat, *Heterocephalus glaber*; (b) the Damaraland mole-rat, *C. damarensis* and (c) the Mashona mole-rat, *C. darlingi*. Concentrations of plasma LH (means ± SEM) in reproductive (Br) and non-reproductive (NBR) female and male mole-rats before (open bars) and 20 min after (filled bars) a single sub-cutaneous injection of GnRH. * $p <$ 0.05, in all cases comparisons of individual means were made using Mann–Whitney U-tests (adapted from Faulkes *et al.*, 1990c, 1991; Bennett *et al.*, 1996b, 1997).

While non-breeding female naked mole-rats in their colonies are normally totally suppressed, there may be some differences among individuals in the extent of their suppression. Westlin *et al.* (1994) demonstrated that when the breeding female is very close to parturition, there was not uniformity in the degree of suppression of the non-breeding females. They suggested that this may be as a result of the queen being unable to direct aggressive 'shoving' behaviour as frequently or as easily as when she is more mobile (Lacey and Sherman, 1991). Westlin *et al.* (1994), however, only examined the females at this one phase in the reproductive cycle of the female. Recently, Van der Westhuizen (1997) examined the basal concentrations of LH in entire colonies during early, mid and late periods of the pregnancy of the queen and found no significant difference in the LH concentrations of non-breeding females at different phases of the queen's cycle.

Van der Westhuizen did, however, find considerable variation between the non-breeding females in the colonies and found a marked correlation in basal LH concentrations with the age of the animal. Older animals tended to exhibit higher bioactive LH concentrations and show a greater response to an exogenous GnRH challenge.

The findings of Westlin *et al.* (1994) and Van der Westhuizen (1997), and reports by Jarvis (1991) of the occurrence of perforate non-breeding females within the colony, illustrate that, amongst the non-breeding females, there are differential levels of suppression, related to the age of the female. O'Riain (1996) and Clarke and Faulkes (1997) have shown quite clearly that queen succession is usually from the older females in the colony. Van der Westhuizen (1997) also documents a succession event in which one of the oldest and largest non-breeders became the new queen.

As with the naked mole-rat, in the Damaraland mole-rat a socially-induced suppression of reproductive physiology is apparent in non-reproductive females. The latter have lower concentrations of basal LH, and reduced sensitivity of the pituitary to exogenous GnRH, when compared with reproductive females (Bennett *et al.*, 1993b; Figure 6.4b). Thus, neuroendocrine suppression mechanisms at the level of the pituitary appear similar in the two species. However, at the level of the ovary, the mechanism of suppression seems to be different, even though the mechanisms controlling ovulation appear to be similar. Both the naked mole-rat and the Damaraland mole-rat are spontaneous ovulators although some evidence suggests that other cryptomids may be induced ovulators requiring the presence of a male and/or mating to trigger ovulation (see Section 5.3; A C. Spinks,

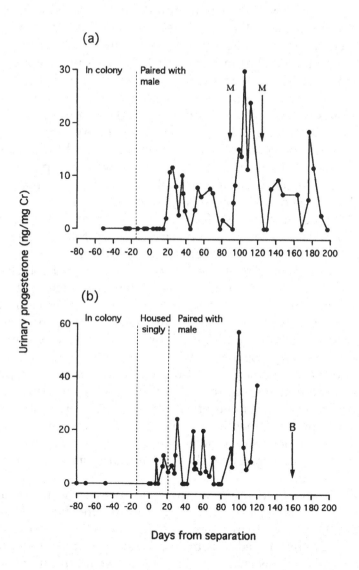

Figure 6.5. Urinary progesterone profiles for two non-breeding female naked mole-rats removed from their colonies and (a) paired with a male straight away; (b) housed singly for six weeks before pairing with a male. The rapid commencement of ovarian cyclicity after separation can be clearly seen, with or without the presence of a male. The peaks in urinary progesterone correspond the the luteal phase of the ovarian cycle. Observations of mating (M) were coincident with the presumed follicular phase of the cycle, when progesterone was low. Female 29 (b) underwent one cycle of around 30 days after pairing with a male, then conceived at the next period of oestrus around day 80. A litter was born about 80 days later (B) (adapted from Faulkes *et al.*, 1990a).

unpublished data; A.J. Molteno, unpubl. data). In contrast to the naked mole-rat, the ovaries of non-reproductive female Damaraland mole-rats show some follicular development, and contain many luteinised secondary and tertiary follicles (Bennett, 1988; Bennett et al., 1994c). A histological examination of 14 females in a colony of 25 mole-rats revealed that 13 of the 14 females were nulliparous. The ovaries of these non-reproductive females were smooth in outline, lacked mature Graafian follicles and were packed in the outer cortex with primordial follicles. Primary and secondary follicles were present, in which the zona pellucida of the ova could be clearly seen. The most distinct feature in these ovaries was the presence of characteristic luteinised unruptured follicles which dominated the cortex and medulla of the ovary in some females.

The luteinised unruptured follicles result from the premature luteinisation of late secondary and early tertiary follicles (normally follicles only undergo luteinisation after ovulation). The *membrana granulosa* and *theca interna* of the luteinised unruptured follicles are packed with ovoid luteinising cells. Non-reproductive females typically possess measurable concentrations of progesterone in both the plasma and urine, which may be attributed to its production in the luteinised unruptured follicles. The presence of the enzyme 3 β hydroxysteroid dehydrogenase, which converts pregnenolone to progesterone via a two-stage reaction, in the luteinised unruptured follicles of these females, suggests that they are the structures which produce progesterone (Bennett et al., 1994c). Bennett et al. (1990) have shown in the aforementioned colony of 25 mole-rats containing 14 females that an increased number of luteinised unruptured follicles occurred with age, suggesting that luteinised unruptured follicles accumulate over time.

Laboratory and field observations confirm that the Damaraland mole-rat, like the naked mole-rat, is capable of breeding throughout the year. Although there are also measurable progesterone concentrations in non-breeding females, sexual activity in the form of solicitation and lordosis is absent. Progesterone concentrations in the reproductive female are, as would be expected, markedly higher in pregnant animals (63 nmols/mmol Cr) but still measurable in non-pregnant breeding females (11 nmols/mmol Cr). Interestingly, the mean concentration of progesterone in non-reproductives is similar (10.7 nmols/mmol Cr). It is possible that the progesterone produced by the luteinised unruptured follicles may result in the non-reproductive females experiencing a 'pseudopregnant' condition that prevents ovulation. Negative feedback action of the progesterone at the hypothalamus and/or pituitary could suppress gonadotrophin release in a manner similar to that of the contraceptive pill (Bennett *et al.*, 1993b). An alternative explanation may be that the luteinised unruptured follicles are formed instead of normal follicles because of an inadequate production and subsequent secretion of LH from the pituitary gland.

In direct contrast to the naked mole-rat, there is no replacement of breeding females from within a colony of Damaraland mole-rats if the queen is not present, e.g. if she dies. Both laboratory and field studies have shown that on the death or experimental removal of one of the founding reproductive animals in the colony, breeding ceases (Jarvis and Bennett, 1993; Rickard and Bennett, 1997). Recrudescence of sexual activity requires the addition of an unfamiliar male to the colony in the case of queen removal, or an unfamiliar female in the case of reproductive male removal (Rickard and Bennett, 1997). Genetic and recapture data from wild colonies (Jarvis and Bennett, 1993; Jarvis *et al.*, 1994; Faulkes *at al.*, 1997a) and behavioural data from laboratory pairings (Bennett *et al.* 1996b) provide irrefutable evidence of the avoidance of breeding with familiar conspecifics and hence incest avoidance. In the laboratory, non-sibling pairings (genetically unrelated and unfamiliar) result in copulatory activity within minutes of animals being placed together and conception occurs. In contrast, pairing of siblings (genetically related and familiar) fails to show any sexual activity and offspring are never produced.

In the laboratory, queenless colonies have maintained a state of sexual quiescence for several years, until an unfamiliar male was introduced. Interestingly, in these colonies the progesterone concentrations in non-reproductive females are significantly lower than the pre-removal values (Bennett *et al.*, 1996b; Figure 6.6). This drop in progesterone concentration may partly be due to release from the reproductive suppression imposed by the queen, and hence a reduction in the number of the luteinised unruptured follicles in the ovary. Females in these colonies also show elevated concentrations of plasma LH and an increased response to administration of exogenous GnRH. Such findings indicate that in the absence of the breeding queen, the physiological component of suppression is lifted, but the incest taboo of the Damaraland mole-rat ensures that breeding does not occur between related colony members (Figure 6.6). However, although the response to a single exogenous GnRH challenge is higher in non-reproductives housed separately from the breeder, it does not resemble that of a breeder (Figure 6.6a). It has

(a)

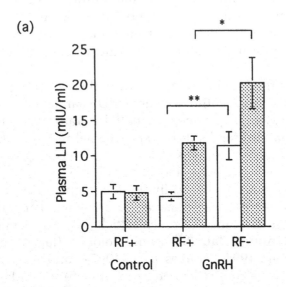

(b)

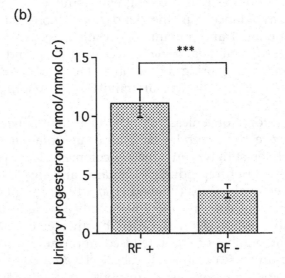

Figure 6.6. (a) Concentrations of bioactive LH (mean ± S.E.M.) in non-reproductive female *C. damarensis* before (open bars) and 20 min after (filled bars) a single injection of 200 μl of saline, a single injection of 2 μg GnRH in 200 μl of saline in non-reproductives in the presence of the queen (RF+), and a single injection of 2 μg GnRH in 200 μl saline in non-reproductives in the absence of a queen for three years (RF-); (b) Concentrations of urinary progesterone in non-breeders where the breeding queen is present (RF+), compared with non-breeders where the queen has been absent for three years (RF-). Adapted from Bennett *et al.*, 1993b, 1996b.

been suggested by Bennett *et al.* (1996b) that the presence of an unfamiliar and genetically unrelated male is the final cue needed to raise concentrations of plasma LH to those found in reproductive females and is what would be expected of animals which have a strong inhibition to incest.

6.4 REPRODUCTIVE SUPPRESSION IN OTHER SPECIES OF BATHYERGID MOLE-RATS

Apart from the naked mole-rat, all species so far investigated in the *Cryptomys* genus appear to live in social groups of parents and at least one litter of offspring (see Chapter 1 for information on group sizes in the various species). In all these social bathyergids, a reproductive division of labour operates, and breeding is monopolised by a single female (except in rare exceptions), and one or a small number of males. But how far does the incidence of physiological suppression of reproduction described for naked and Damaraland mole-rats extend to the other species? In these other species is incest avoidance sufficient to maintain the reproductive division, and how do such observations correlate with the degree of sociality? Such sudies are now progressing and the extent of our knowledge is summarised in Table 6.1.

Burda (1995) has found evidence for incest avoidance in a Zambian mole-rat (probably *C. amatus*). He found that if siblings were removed from their colony in captivity and housed separately for more than 14 days, they would mate if they were subsequently paired together, suggesting that it is cues based on familiarity, rather than genetic recognition of kin, that are used to avoid inbreeding. Burda (1995) has also suggested that non-reproductive females in colonies of another Zambian *Cryptomys* species are not reproductively suppressed by the breeding female either behaviourally or physiologically. He suggests that they do not mate with their fathers or brothers simply because they are avoiding inbreeding but at present there are no endocrinological data to back this claim and prove that physiological suppression does not also occur. However, this suggestion by Burda is supported by studies of another social cryptomid, the Mashona mole-rat, *C. darlingi*. This species shows no significant difference in circulating basal and GnRH stimulated LH levels between the reproductive and non-reproductive animals of either sex (Bennett *et al.*, 1997; Figure 6.4c). Thus, neither males nor females appear to be physiologically

Table 6.1. Summary of social and reproductive parameters that illustrate the range of suppression mechanisms within and between cooperatively breeding species of African mole-rats. Reproductive control is the greatest in naked mole-rats, *H. glaber*, where there is no incest avoidance and large constraints on dispersal, and shows a decreasing trend in the other species where incest avoidance is sufficient to maintain singular breeding within colonies. * denotes a seasonal effect on reproduction in the common mole-rat, *C. h. hottentotus.*

Social/reproductive parameter			*H. glaber*	*C. damarensis*	*C. darlingi*	*C. h. hottentotus*
Social system			Eusocial	Eusocial	Colonial	Colonial
Reproductive division of labour			Yes	Yes	Yes	Yes
Behavioural division of labour			Yes	Yes	No	No
Incest avoidance			No	Yes	Yes	Yes
F E M A L E S	Pituitary:	Basal LH differences	Yes	Yes	No	No*
		Difference in response to exogenous GnRH	Yes	Yes	No	No*
	Ovary:	Hormonal and structural differences	Yes	Yes	?	?
M A L E S	Pituitary:	Basal LH differences	Yes	No	No	No
		Difference in response to exogenous GnRH	Yes	No	No	No
	Testis:	Differences in testosterone	Yes	No	?	?
		Differences in sperm	Yes	No	No	No

suppressed, and the reproductive skew observed in colonies appears to be maintained entirely through inhibition of reproductive behaviour, possibly as a result of inbreeding avoidance.

Similarly, in the common mole-rat, a physiological block to reproduction in non-breeders of both sexes is absent. The common mole-rat occurs in the winter rainfall region of the Cape Province, and is unusual in that it is the only colonial bathyergid discovered so far that exhibits a well defined breeding season. Interestingly, males show no seasonal difference in testicular activity, spermatogenesis and sperm quality (motility and normal morphology; Spinks *et al.*, 1997). The maintenance of reproductive activity in the non-breeding period may be of adaptive significance, since it coincides with the period when maximal dispersal takes place. It has been suggested by Spinks *et al.* (1997) that dispersing males being reproductively active may facilitate intersexual recognition, and promote pair bond formation prior to the breeding season. Similarly, none of the aforementioned reproductive parameters, together with plasma LH levels, differed between breeding and non-breeding males (Spinks, 1998). Likewise in females, there was no effect of either season or reproductive status on ovarian histological or anatomical parameters, other than differences relating to pregnancy in the breeding females. However, there is some evidence to suggest that there may be seasonal reductions in the responses to administration of exogenous GnRH among females. Thus, during the non-breeding season, females may be protected against conceiving and giving birth at an inappropriate time of the year (Spinks, 1998).

6.5 PROXIMATE CUES AND REPRODUCTIVE SUPPRESSION IN MOLE-RATS: A REPRODUCTIVE DICTATORSHIP

In the preceding sections we have seen that, aside from incest avoidance in the cryptomids, there is also a physiological component to the maintenance of the reproductive division of labour in both sexes of naked mole-rats, and in females of the Damaraland mole-rat. But what are the social cues responsible for bringing about this suppression of reproduction? The importance of primer pheromones (chemical signals released by conspecifics that elicit a physiological response in the recipient animal) in modulating reproduction in microtine rodents is well known (Vandenbergh, 1988). For example, adult female bank voles, *Clethrionomys glareolus* (Kruczek and Marchlewska-Koj, 1986) produce a urinary pheromone which delays puberty in juvenile females. Similarly, in microtine rodents occurring in extended family groups, suppressing pheromones released by the adults may function to reduce reproductive competition by encouraging dispersal of offspring or preventing inbreeding (Bazli *et al.*, 1977). It was initially suggested that pheromones in the urine of the breeding female may have suppressed reproduction in naked mole-rats (Jarvis, 1981). However, Faulkes and Abbott (1993) showed that primer pheromones from urine, or other secretions contained in soiled bedding or litter, do not play a major role in the suppression of reproduction in non-breeding female or male naked mole-rats. The addition of soiled bedding to an experimental group of females, housed separately or paired with males, failed to prevent reproductive activation. Both control and bedding transfer groups of singly housed females commenced ovarian cyclicity. The time taken for the testosterone titre of a non-breeding male, following removal from his colony, to resemble that of a breeding male, was not significantly different between the control and bedding transfer groups, i.e. approximately five days. Instead of pheromones, a behavioural component involving direct contact with the breeding queen has been suggested for suppression, possibly involving either overt or subtle harassment, or both (Faulkes and Abbott, 1993, 1997; Smith *et al.*, 1997). How these cues are processed and subsequently affect the neuroendocrine system of the hypothalmus remains a mystery, although there does not appear to be a causal link between elevated levels of the stress hormone cortisol and suppressed reproduction in naked mole-rats (Faulkes and Abbott, 1997).

The breeding female naked mole-rat is typically the most dominant animal in the colony, and rival challengers for breeding status are usually attacked and killed by the queen. However, most of the time naked mole-rat colonies are harmonious societies with little evidence of overt aggression. One of the few agonistic behaviours that is seen on a regular basis is 'shoving'. The queen initiates significantly more shoving events than any other colony member. These bouts of shoving involve the subordinate animal being forced backwards down the tunnel violently by the queen (Jarvis, 1991; Lacey *et al.*, 1991; Reeve and Sherman, 1991; Reeve, 1992; Jacobs and Jarvis, 1996; O'Riain, 1996). Despite the high degree of inbreeding normally seen in wild colonies (Reeve *et al.*, 1990; Faulkes *et al.*, 1990b, 1997b), evidence of nepotism has been found in colonies of animals having mixed relatedness (Reeve and Sherman, 1991; Reeve, 1992). The breeding female was found to initiate significantly more shoving interactions with colony members that were less related to her. The number of shoving interactions was also found to correlate positively with body size of the recipient animal. Reeve and Sherman (1991) put forward two hypotheses to explain shoving, neither exclusive of the other. The first hypothesis proposes that the breeding female shoves colony mates to indicate her reproductive dominance and her willingness to fight, and this reduces the threat of the colony members from challenging her for her breeding status. Reeve & Sherman (1991) propose that in colonies of mixed relatedness (which could occur in the wild as a result of male dispersal), the mole-rats of lower relatedness would be shoved more because their ascendance to a position of reproduction would incur the most severe inclusive fitness loss to the reigning queen, and they have more potential fitness benefits to gain from direct reproduction. It is further postulated that the larger animals are shoved more frequently because they represent a greater threat to her reign. The second hypothesis, the 'activity incitation hypothesis', suggests that shoving increases activity in so called 'lazy' workers. This hypothesis assumes that a worker's activity levels vary inversely with its relatedness to the breeding female. This assumption is reasonable if the activity level of the mole-rat represents a balance between the kin selective benefits of colony maintenance activities and personal fitness costs of such efforts. Reeve and Sherman (1991) also suggest that shoving in breeding female naked mole-rats is analogous to queen aggression in social insects.

Jacobs and Jarvis (1996), however, found no evidence that the breeding female's shove rate and the amount of colony maintenance activities performed by non-breeders are influenced by relatedness. They found that body size was a better predictor of work rate and shove rate than relatedness, i.e. large animals, regardless of their relatedness to the queen, were shoved more and worked less. They also found that mole-rats were more likely to be shoved when already active than when resting in the nest. Their results support the first of the two hypotheses put forward by Reeve and Sherman, namely that of reproductive suppression of the non-breeders, rather than the activity incitation hypothesis. Shoving may thus be one behavioural component in initiating and maintaining physiological reproductive suppression in non-breeding male and female naked mole-rats (Faulkes and Abbott, 1993; Jacobs and Jarvis, 1996; O'Riain, 1996; Clarke and Faulkes, 1997, 1998).

While studies of the proximate factors involved in reproductive suppression in naked mole-rats are fairly well advanced, little is known about what social cues may be involved in the physiological suppression of females in the Damaraland mole-rat. Certainly, the agonistic behaviour, including shoving, so characteristic of the queen in naked mole-rats appears to be absent in Damaraland mole-rats, and overt aggression is rare. This tends to argue against dominance-based behavioural cues being involved in suppression, although breeding females are the most dominant of their sex and normally near the top of the colony hierarchy below the breeding male (Jacobs *et al.*, 1991). Further work is now required to elucidate the social control mechanisms that may be operating in this species, and in particular, the potential role that pheromones may play.

6.6 MARMOSETS, MONGOOSES AND MEERKATS

One of the main aims of optimal reproductive skew theory is to develop a unified, cross taxon approach to understanding the evolution and maintenance of cooperative breeding systems. Therefore, it is of value to make comparisons between sociality, ecology and the mechanisms and degree of reproductive suppression among different mammalian taxa. With regard to the physiological mechanisms and degree of reproductive suppression, studies are well advanced in some herpestid species like the dwarf mongoose, *Helogale parvula* (see Creel and Waser, 1997 for review), canids (reviewed in Moehlman and Hofer, 1997 and Asa, 1997), and callitrichid primates like the common marmoset, *Callithrix jacchus* (Abbott *et al.*, 1997). In this section, we will briefly review some of the principal findings from studies of these other cooperatively breeding species, for comparison with the African mole-rats.

Marmosets and tamarins in the family Callitrichidae, like the Bathyergidae, exhibit a diversity of proximate mechanisms that are involved in maintaining singular cooperatively breeding systems, i.e. a single breeding pair and helpers. Species within the family are characterised by fairly small territorial groups of three to 15 individuals, that have a wide distribution throughout the primary and secondary forests of South America (see French, 1997 for review). Common marmoset groups of up to around 15 individuals in the wild are primarily extended families, but unrelated immigrants may also be found. In both wild and captive colonies, reproduction is restricted to a single dominant breeding female, while the non-breeders act as helpers by assisting with the carrying of the breeding female's twin offspring (even at birth they constitute a burden, as their body mass is some 20% of an adult). As with other cooperative breeders, reproductive suppression in male marmosets appears to be predominantly behavioural rather than physiological. Non-breeding male marmosets in a family group exhibit little sexual behaviour, although they will mate with unfamiliar females. As these males are offspring of the breeding male, the lack of sexual behaviour presumably results from an incest avoidance mechanism. However, in captive 'peer' groups of common marmosets consisting of unrelated adults, the dominant breeding female mates mainly with the rank one dominant male. Subordinate males are also seen to mate, but at a lower frequency, and usually such behaviour brings them into conflict with the dominant male, who may attack them, or terminate their copulations by harassment and displacement (Abbott, 1984). Subordinate males do have reduced plasma concentrations of LH and testosterone, but whether these deficiencies result in impaired spermatogenesis and quality and quantity of mature spermatozoa remains unclear.

Common marmoset females are spontaneous ovulators, and studies in captivity have revealed that ovarian cyclicity and ovulation is blocked in non-breeding females (Abbott, 1984). As with both Damaraland and naked mole-rats, non-breeders had reduced concentrations of plasma LH as a result of changes in secretion of, or sensitivity to, hypothalamic GnRH. This hypogonadotrophism is reversible if females are removed from their social groups, or remain in their social groups and are given GnRH replacement therapy. However, from direct measurement of GnRH in the hypothalamus, it appears that GnRH release does not differ between non-breeding females and breeding females in the follicular phase of their ovarian cycle. This suggests that rather than GnRH release being inhibited in the hypothalamus, other factors, like a reduction in the sensitivity of the pituitary to GnRH, may be implicated (Abbott *et al.*, 1997). The underlying neuroendocrine mechanisms that suppress pituitary LH in female common marmosets appear to bear some similarities to the seasonal suppression of estrous in the ewe, and lactational amenorrhoea in rats. Part of this mechanism is an increased sensitivity to the feedback effects of oestradiol in non-breeders, and a separate pathway involving endogenous opioid peptides (Abbott *et al.*, 1997). The link between these two factors and the impairment of GnRH action within the brain remains to be established.

Reduced body weight and elevated levels of the hormones prolactin and cortisol have all been implicated in impaired ovarian function, and hyperprolactinemia and hypercortisolemia have also been associated with social stress and subordinate social status in females. However, none of these factors are implicated in social suppression of reproduction in common marmoset monkeys. Instead, it appears that associative learning of olfactory and visual cues from the dominant female, by subordinates, results in a psychological conditioning component to the neural and neuroendocrine mechanisms that suppress reproduction in non-breeding females (Abbott *et al.*, 1997). It is possible that a similar proximate mechanism is involved in mediating suppression in naked mole-rats. During the establishment of a queen in a colony when agonistic interactions are high, associative learning of the dominant queen's presence and intimidation may lead to subtle behavioural cues from the queen becoming conditioned stimuli, leading to continued suppression of reproduction. Such stimuli could include individual-specific odours (but not a ubiquitous primer pheromone), vocalisations of the queen and tactile stimuli like shoving.

Another important mammalian group for comparative analysis of reproductive control mechanisms in cooperative breeders are the various mongoose species in the carnivore family Herpestidae. As with the Bathyergidae, mongooses have a wide distribution across sub-Saharan Africa, and although most species are solitary, both meerkats, *Suricata suricatta* and dwarf mongooses, *Helogale parvula* are cooperative breeders with a reproductive division of labour (Rasa, 1973; Creel and Waser, 1997; Doolan and Macdonald, 1997). At least three species of mongoose are known to be induced ovulators, and endocrine investigations of dwarf mongooses have revealed that non-breeding females had significantly lower concentrations of urinary estradiol and lower mating rates. As oestrogen priming prior to mating is needed for induction of ovulation, the implication is that even if some non-breeding subordinate females mate, they probably will not ovulate (Creel *et al.*, 1992; Creel and Waser, 1997). Among non-breeding subordinate male mongooses, androgen levels are not different from breeding males, and thus dwarf mongooses seem to follow the typical pattern of most male mammalian cooperative breeders, with proximate control of suppression being behavioural rather than physiological (Creel *et al.*, 1992).

In conclusion, it is clear that among mammalian species, there are some common features in the proximate mechanisms mediating socially-induced reproductive suppression, but there are also species differences, even within close taxonomic groups. The same is true for the ultimate factors involved in the evolution and maintenance of cooperative breeding. In the Herpestidae, predator avoidance by guarding behaviour may be a principal factor (Doolan and Macdonald, 1997; Creel and Waser, 1997). In the Callitrichidae the large energetic burden of carrying twin offspring may be important in the evolution of helper behaviour. Adult marmosets and tamarins are small for primates, weighing from 100 to 700g. At birth, a twin litter weighs approximately 16 to 20% of the adult body weight, and all group members may participate in carrying infants, although there is much variation in individual participation and the benefits of this helping behaviour remain unclear (French, 1997; Tardif, 1997). In the Bathyergidae, the subterranean niche imposes a different set of problems, and here ecological constraints, in the form of rainfall and distribution patterns of food, are thought to be important determinants of sociality. These factors will be discussed fully in Chapter 8.

6.7 THE ADAPTIVE SIGNIFICANCE OF REPRODUCTIVE SUPPRESSION: THEORIES OF OPTIMAL REPRODUCTIVE SKEW

The African mole-rats of the family Bathyergidae represent a unique model system to test hypotheses on social evolution. They exhibit a range of social behaviours and mechanisms to restrict reproduction to a single breeding female, and occur in a wide range of climates (Jarvis *et al.*, 1994; Bennett *et al.*, 1997). Not only can we look for correlations between social, ecological and phylogenetic factors within the family (see Chapter 8), but comparisons can also be made across taxa with other social vertebrates and invertebrates. A unified approach to making such comparisons has been developed in theories of reproductive skew and lifetime reproductive success (Vehrencamp, 1983; Keller and Reeve, 1994; Sherman *et al.* 1995). It is sometimes useful to consider cooperatively breeding animal societies as a continuum of social systems that differ in the way that lifetime reproductive success is distributed among group members (Sherman *et al.*, 1995). In 'low skew' societies, such as the solitary bathyergids, all individuals have an equal chance of reproduction. In 'high skew' societies, reproduction is limited to one or a small number of individuals of each sex. Because reproduction is confined to a single female, all social bathyergids have a high skew. However, as pointed out by Sherman *et al.* (1995), it is possible to have social groups with the same skew but with a different lifetime reproductive success. It is here that differences may become apparent in the social bathyergids once sufficient field information is available on colony dynamics of species in mesic areas.

Skew theory predicts that the reproductive skew and lifetime reproductive success in a society will be influenced by four parameters: (i) a measure of amount of ecological constraint on independent breeding, or a measure of the success of a subordinate breeding solitarily; (ii) a measure of the group's productivity if the subordinate stays within the group and cooperates; (iii) the genetic relatedness among the members of a social group, and (iv) the fighting ability of the subordinate, or how likely it would be to fight and depose a dominant breeder (Keller and Reeve, 1994). A critical assumption of skew theory is that dominant members of the society control reproduction in the subordinates. Clearly, in a 'closed' society such as we find in subterranean mole-rats, where immigration and emigration are quite rare events, incest avoidance mechanisms may be sufficicent to effectively maintain reproductive suppression. Skew theory also postulates that if dominants benefit from the presence of subordinates, it may pay the dominant to allow some subordinates to breed, as an inducement to stay in the group and cooperate peacefully. These 'staying incentives' (reproductive inducements that prevent subordinates leaving) are predicted to decrease as the ecological contraints increase, because the chances of successful independent breeding for subordinates dispersing are low.

When considering reproductive responses to ecological constraints in social mole-rats the picture is complicated by an additional variable. This is the fundamental disparity in mating systems that separates the two genera of social bathyergids, i.e. the inbred naked mole-rat and the outbred cryptomids. Among outbreeding social mole-rats that inhabit mesic habitats, opportunities for dispersal and of becoming reproductives in new colonies are potentially high. In these situations, one would predict that the staying incentives offered by the breeders to subordinates would be higher than in areas with strong environmental constraints. In other words, the breeder may 'allow' some subordinate reproduction in order to stop them dispersing from the colony, for example if a foreign animal joins the colony (incest avoidance would prevent subordinate reproduction at other times). In these species there is no evidence that the reproductives are in fact controlling reproduction in their non-breeding offspring (*sensu* Vehrencamp, 1983; Keller and Reeve, 1994), except perhaps by preventing foreign animals from joining the colony. We await with interest the results of ongoing research on the population dynamics of the common mole-rat (A.C. Spinks, unpubl.) to see if there is evidence of the use of staying incentives by the reproductives.

In marked contrast to the mesic species, the naked mole-rat inhabits areas in which the ecological constraints to successful dispersal are high, and therefore, theoretically, the reproductive female need offer few incentives to the non-breeders to stay. Naked mole-rats, however, show the most extreme form of reproductive suppression found in the Bathyergidae, in which there is physiological suppression of both sexes of non-reproductives. This may be a response to this species being a facultative inbreeder. The absence of incest avoidance necessitates the evolution of a stringent reproductive control and consequently the marked degree of reproductive suppression exhibited in this species. A strong reproductive suppression may also be necessary to counter the otherwise high incentives a non-breeder would have to attain the reproductive position in a colony where the lifetime reproductive success of non-breeders is close to zero. Despite this reproductive suppression, challenges to the reproductive female do sometimes occur. There is also often intense conflict during contests for succession to breeding position after the death of the breeding female (Jarvis, 1991; Lacey and Sherman, 1991).

The Damaraland mole-rat is intriguing as its mechanisms of reproductive suppression involve both incest avoidance and a physiological block. Although incest avoidance alone (behavioural suppression) appears to be operative in the non-reproductive males, the non-reproductive females have both behavioural and physiological suppression preventing them from breeding (Bennett *et al.*, 1996b). This additional form of reproductive control at first appears unnecessary in a colony consisting entirely of siblings and their parents where incest avoidance alone should suffice. However, in the extremely arid regions where Damaraland mole-rats occur, there would be strong advantages to all members of small colonies in retaining their workforce if a breeder dies. Indeed, in these small colonies there is often a high turnover of reproductives, presumably because they too have to participate in high-risk tasks in the colony because the workforce is small (Jarvis *et al.*, 1998). In instances such as this, where a new unrelated replacement reproductive enters the colony, incest avoidance would no longer be operative and it would be advantageous for the reproductive to actively suppress reproduction in the workers. Under the drought conditions that are often a characteristic of the Damaraland mole-rat habitat, the breeder would probably not have to offer staying incentives to the non-breeders, and it would be interesting to see if the situation changed in good rainfall years when environmental constraints to dispersal were lower. As mentioned earlier, this mechanism of replacement of a reproductive does not appear to occur in large colonies and here the colony fragments on the death of a reproductive. This fragmentation may be delayed for more than a year until conditions are optimal for dispersal. On a few occasions foreign 'floater' males joining a large colony have been recorded in the field (N.C. Bennett and J.U.M. Jarvis, unpubl.). In these instances physiological suppression of the non-breeding females in the colony would be in the interests of the breeding pair.

The field studies on both the naked and Damaraland mole-rats have shown that there is a very low lifetime reproductive success for the majority of the colony members. Thus, less than 0.1% of naked mole-rats and less than 10% of Damaraland mole-rats successfully disperse, and even then, many of these dispersers never successfully found a colony (Jarvis *et al.*, 1994). We predict that this situation will be different in the social mole-rats inhabiting mesic environments and that lifetime reproductive success will be much higher (Bennett *et al.*, 1997).

It would therefore seem that, probably as a consequence of early divergence in the lineages of the two social genera (Allard and Honeycutt, 1992; Faulkes *et al.*, 1997a), two very different mating systems have evolved within the family Bathyergidae. These differences have had an impact on the responses of the social species to the constraints of aridity and the way in which monopoly of reproduction has been achieved and is maintained. In colonies of the inbred naked mole-rat where opportunities to disperse are few, closely related individuals are forced to remain in the colony and help. The long 'reign' of the breeding female in a colony means that the lifetime reproductive success of the non-breeders is extremely low and, unless the breeding female is strongly suppressing reproduction in the non-breeders, it would be in their best interest to challenge her position and become the new breeder in the colony. Damaraland mole-rats, because of the constraints of the mating system of their lineage, have responded somewhat differently to severe environmental constraints. They have retained incest avoidance in both sexes but have added a physiological component to the reproductive control of the females. This then prevents the collapse of reproductive control if a new breeder joins a small colony on the death of one of the breeding pair and thereby ensures that the workforce is retained in these highly vulnerable small colonies.

Snowdon (1996) has proposed that there are essentially two models which regulate reproductive suppression in cooperative breeding animals. These mechanisms are (i) the dominant control model and (ii) the self restraint model.

The dominant control model states that reproductive suppression is under the control of the dominant breeding animal. This mechanism requires that reproductive suppression is imposed by aggression directed at the subordinate females by the breeding female. In this model it is the presence of the dominant female that prevents other females attempting to breed and the absence of the dominant female can result in a previously subordinate female attaining breeding status from within. This situation mirrors the finding in the highly inbred naked mole-rat, where aggression in the form of physical shoving is apparently responsible for the suppression imposed upon non-breeding subordinate females. Likewise, removal of the queen is usually met by succession from within the colony. The role played by outbreeders still needs to be clarified from field studies (O'Riain *et al.*, 1996).

The self restraint model does not involve aggression directed towards the non-reproductive females. In contrast, the subordinate female can control her interactions with the dominant female. This is the basic model found in the cryptomids that employ incest avoidance, although it is possible that the reproductives direct aggression, not at subordinate females, but at any foreign animals attempting to join the colony. Despite this possibility it appears that only the Damaraland mole-rat clearly shows features of both models.

Both naked mole-rats and the cryptomids are very xenophobic to non-colony members. It appears that recognition of colony members in all social mole-rats is based on familiarity and colony odour and does not have a genetic basis, although, because colonies are extended families, colony members are almost always close kin (O'Riain and Jarvis, 1997; Burda, 1995).

6.8 SOME CONCLUDING COMMENTS

Incest avoidance is the rule in the order Rodentia, and indeed in most animals (Blouin and Blouin, 1988), and appears to be a characteristic in all species of social mole-rats in the genus *Cryptomys* (Bennett *et al.*, 1997). However, the eusocial naked mole-rat, a monotypic species with a lineage that is divergent from the other Bathyergidae, is a facultative inbreeder (Jarvis, 1991) with only limited opportunities to outbreed (O'Riain *et al.*, 1996). The evolutionary pressures that led to this different lifestyle in the naked mole-rat are unknown, although presumably at some time in their history opportunities to outbreed were severely curtailed. This dichotomy in mating systems has resulted in divergent mechanisms used to monopolise reproduction in colonies of the two social genera, both of which have a high reproductive skew. All the cryptomids have colonies that are composed of an outbred reproductive pair and their offspring and here incest avoidance will usually ensure that none of the offspring will breed within their natal colony. Some of the species of *Cryptomys* occur in mesic areas and here it is predicted that the lifetime reproductive success of the non-breeders will be higher than the species occurring in arid regions, e.g. the Damaraland mole-rat. In the inbred naked mole-rat, where incest avoidance is not operative and there is a very low lifetime reproductive success for the non-breeders, there is a strong incentive for the non-breeders to compete for the reproductive position. Here, the breeding female uses dominance interactions to physiologically suppress the non-reproductive males and females in the colony. Particularly interesting are the responses of these two lineages to the severe environmental constraints to dispersal experienced in arid parts of Africa, where the lifetime reproductive success of non-breeders is very low. The naked mole-rat inbreeds and imposes a physiological block to breeding in subordinates mediated through dominance interactions. The Damaraland mole-rat retains the incest avoidance of the cryptomid lineage, and only exhibits a physiological component to reproductive suppression in the non-breeding females, possibly to ensure that the reproductive division of labour and colony stability is maintained in instances when a foreign male enters the colony, and incest avoidance on its own is not sufficient.

These different mating systems and mechanisms of reproductive suppression may well provide insights into which selective pressures are central to the evolution of sociality and reproductive suppression in the Bathyergidae, and which are special features attributable to different lineages within the family.

Chapter 7

The genetic structure of mole-rat populations

7.1 GENETIC RELATIONSHIPS IN SUBTERRANEAN MAMMALS AND COOPERATIVE BREEDERS

Both the constraints of the underground niche, which by its very nature limits gene flow, and the cooperatively breeding social strategies of the Bathyergidae will influence the genetic composition and relationships in their colonies and populations. Apart from studies of the Bathyergidae described in this chapter, research on the population genetics of subterranean mammals has focused mainly on a myomorph rodent, the solitary dwelling blind mole-rats in the genus *Spalax*, found across the Middle East. This taxon exemplifies the consequences of living underground on the subsequent genetic structuring of populations. In *Spalax*, limited gene flow has led to many local forms which are difficult to identify, because convergent evolution and adaptation to the subterranean niche has led to extreme morphological similarities. As with the Bathyergidae, the taxonomy and systematics are in need of revision. Despite there being more than 30 extant chromosomal forms having different karyotypes, currently only one genus and eight superspecies are recognised (Nevo et al., 1995). These chromosomal species of *Spalax*, which have arisen from Robertsonian translocations during meiosis, resulting in a speciation trend from low to high chromosome numbers, almost certainly represent different biological species. They occur over a large range either allo- or parapatrically, in different climatic regions, with no evidence of exchange of genes across the narrow hybrid zones that exist between some of the chromosomal races (Nevo, 1991; Nevo *et* al., 1993, 1995; Suzuki *et al.*, 1996). Local selection and random fixation of genotypes have given rise to many mitochondrial DNA haplotypes that are unique to a chromosomal species or population (Nevo *et al.*, 1993). The four chromosomal forms of the *Spalax ehrenbergi* superspecies that occur parapatrically in Israel have been shown to be quite divergent. Overall estimates of genetic distances for the entire superspecies, derived from restriction fragment length polymorphism (RFLP) analysis, were high when compared to similar within-species estimates for naked mole-rats, *Heterocephalus glaber*, common mole-rats, *C. h. hottentotus* and the Natal mole-rat, *C. h. natalensis* (Nevo *et al.*, 1993). Similarly, studies of *S. ehrenbergi* and another superspecies within the genus, *S. leucodon*, in Turkey, Egypt and Israel revealed similar trends (Suzuki *et al.*, 1996). Interestingly, across the range of *Spalax* in both Turkey and Israel, chromosome number, heterozygosity and genetic diversity at a number of loci show positive correlations with increasing habitat aridity. It has been suggested that these correlations reflect an adaptive response to environmental aridity stress, although the functional significance of these trends remains unclear (Nevo 1991; Nevo *et al.*, 1993, 1995). Among the Bathyergidae, chromosome numbers range from 2n = 40 in the giant Zambian mole-rat, *Cryptomys mechowi*, which lives in a mesic habitat, to the naked mole-rat (2n = 60) and the Damaraland mole-rat, *C. damarensis* (2n = 74 and/or 78). The two latter species live in the most arid habitats. However, this trend is confounded by the silvery mole-rat, *Heliophobius argenteocinereus*, which is found in mesic areas of high rainfall, but also shares a 2n of 60 with the naked mole-rat (see Chapters 1 and 8).

Apart from effects of stochastic processes that arise from limited gene flow, fluctuating population sizes and genetic drift, another explanation proposed for the genetic patterns seen in subterranean mammals is the 'niche width genetic variation' hypothesis. Nevo (1979) suggests that the reduced genetic variation seen in subterranean mammals compared with small mammalian species living above the ground (Nevo *et al.*, 1990a) results from the narrow subterranean niche which is stable and predictable. However, this idea has been challenged by others who argue for the former stochastic processes being the predominant factor (e.g. Sage *et al.*, 1986).

A second major factor that has the potential to profoundly influence the genetic structure of mole-rat populations is the social structure, particularly in cooperative breeders. In social species, genetic relationships within groups are a major factor in determining the kind of reproductive and behavioural strategies expressed by individuals. In cooperatively breeding species like some of the Bathyergidae, a knowledge of the genetic structure of populations, and in particular genetic relatedness within groups, is crucial if we are to understand the factors involved in the evolution and maintenance of these social systems. The relationships between genetic relatedness and the costs and benefits of cooperative behaviour were first quantified by Hamilton (1964) for social insects, to explain how selection may favour altruistic traits. He developed his hypothesis from the study of bees which exhibit a phenomenon known as haplodiploidy. In this unusual system of sex determination, common to many social Hymenoptera (wasps, ants and bees), females are diploid in the normal way, but males arise from unfertilised eggs and are therefore haploid. Furthermore, because of this haploid state, the sperm of a male are all genetically identical. Therefore, assuming that a female (queen) mates only once (which it is now known does not always occur), all her female offspring will receive an identical set of genes from the haploid father. The mother is diploid, so females will have in common, on average, half of the maternal genes (the normal pattern of inheritance). Thus, sisters will share, on average, three quarters of their genes, while they would only share half with their own offspring. Thus, in terms of fitness benefits, it pays these hymenopteran females to help their mother to produce more sisters rather than for the females to have their own offspring, as an extra 25% of genes will be perpetuated. Although haplodiploidy explains altruistic behaviour in eusocial Hymenoptera, it does not occur in all insects (e.g. eusocial termites are diploid), or in vertebrates. Haplodiploidy is therefore not a mandatory prerequisite for eusociality, but it does serve to emphasise the importance of genetic factors in understanding altruistic behaviour in other species.

In recent years Hamilton's ideas have been extended and incorporated into models of optimal reproductive skew, which have developed to address questions on the evolution of cooperative behaviour in vertebrate societies (Alexander, 1974; Vehrencamp, 1983; Keller and Reeve, 1994). These models attempt to account for the elements of conflict and competition which are usually present in such societies, together with the degree of reproductive skew, i.e. a measure of the numbers of animals of each sex that are reproducing in groups, and their relatedness. In the bathyergid rodents that have varying degrees of sociality and reproductive skew, one would predict changing patterns of relatedness in these different species, and an understanding of intra-group genetic relationships is clearly

essential if we are to understand the factors involved in the evolution
of sociality in the Bathyergidae (see Chapter 8).

Molecular genetic techniques can provide a vast amount of
information about the social and reproductive biology of a
population, without the need to observe animals at length. These
techniques are especially suited to the study of subterranean animals,
which are effectively impossible to observe directly in their natural
habitat. Broadly speaking, such molecular methods fall into two main
categories: sequence analysis, and multilocus genotyping of one form
or another.

Sequence analysis involves determining the sequence of bases at
one or more genetic loci, and then comparing this sequence between
individuals, quantifying any differences that have arisen through
mutations, which accumulate over time. The greater the percentage
sequence differences between individuals, the greater their divergence
time, or the time since they shared common ancestry. From these
data, given certain assumptions, phylogenetic relationships can be
calculated and expressed as a branching tree or cladogram (e.g.
Figures 1.1 and 1.13). Mitochondrial DNA sequence analysis is most
often used for such phylogenetic studies, as it is clonally inherited
through the maternal line (see Harrison, 1989 for review). In this case,
the phylogenies represent the matrilineal history of the gene or locus
(and by inference, the individual or species).

Multilocus approaches to quantifying the genetic makeup of
individuals and populations were made famous in sociobiology by
the early DNA fingerprinting studies of mating behaviour in birds
(e.g. Burke and Bruford, 1987). More recently, minisatellite analysis
has been largely superceded by microsatellite genotyping methods.
These are based on the polymerase chain reaction (PCR) and
therefore need only tiny quantities of sample and are highly specific.
Microsatellite genotyping can provide information about relatedness,
parentage, mating systems and gene flow within and between
populations (see Bruford and Wayne, 1993 for review).

Apart from molecular phylogenetic studies within the family
Bathyergidae, genetic analysis at the intra-specific level has been
carried out in three species of mole-rats. Currently, the population
genetics of the Damaraland mole-rat, the common mole-rat and the
naked mole-rat have been investigated. However, the naked mole-rat
has received by far the most attention to date, and estimates of
genetic relatedness, important in understanding the evolution of
cooperative behaviour, have at present only been performed in this
species (Reeve *et al.*, 1990; Faulkes *et al*, 1990b, 1997b).

7.2 MICRO- AND MACRO-GEOGRAPHIC GENETIC STRUCTURING OF NAKED MOLE-RAT COLONIES

As we have seen in earlier chapters, the reproductive division of labour in colonies of naked mole-rats is manifest as a monopoly of breeding by a dominant female, the 'queen', and one to three males (Jarvis, 1981; Lacey and Sherman, 1991). The remaining colony members of both sexes are reproductively quiescent, but not sterile, and are classified as non-breeders (Jarvis, 1981; Faulkes *et al.*, 1990a, 1991; Lacey and Sherman, 1991). During oestrus, the breeding queen appears to initiate courtship and solicits mating from only the breeding male. If more than one breeding male is present, the queen may mate with the other(s) on numerous occasions during the oestrus period (Jarvis, 1991), giving rise to multi-male paternity and, potentially, the formation of more genetically heterogeneous litters (see Chapter 5).

The subterranean colonies of naked mole-rats exceed all other bathyergid species in their group sizes (commonly 40 to 90, but sometimes over 295 individuals), and total tunnel length within the burrow may exceed 3 km (Brett, 1991b; Chapter 2). Due to the xenophobic nature of naked mole-rats (Lacey and Sherman, 1991) and the high cost of dispersal for individuals or small groups of animals (Lovegrove and Wissel, 1988), these wild colonies are almost completely isolated breeding groups. New colonies are thought to form almost exclusively through fission of existing colonies (Brett, 1991b). The combination of environmental constraints that limit dispersal and the reproductive strategy of the naked mole-rat would be expected to produce colonies that naturally become highly inbred and genetic studies of wild colonies have confirmed this hypothesis (Faulkes *et al.*, 1990b, 1997b; Reeve *et al.*, 1990; Honeycutt *et al.*, 1991b).

A survey of wild colonies from Ethiopia, Somalia and locations in northern and southern Kenya by Faulkes *et al.* (1997b) examined mitochondrial DNA (mtDNA) control region sequence variation in 42 individuals from 15 colonies, together with multi-locus DNA fingerprinting and mtDNA cytochrome-*b* sequence analysis in selected individuals. While different populations showed considerable genetic divergence, individuals within colonies, and even between colonies at certain locations, were genetically almost monomorphic at these loci. They shared the same mtDNA control region haplotype, indicating a recent common maternal ancestor (Figure 7.1).

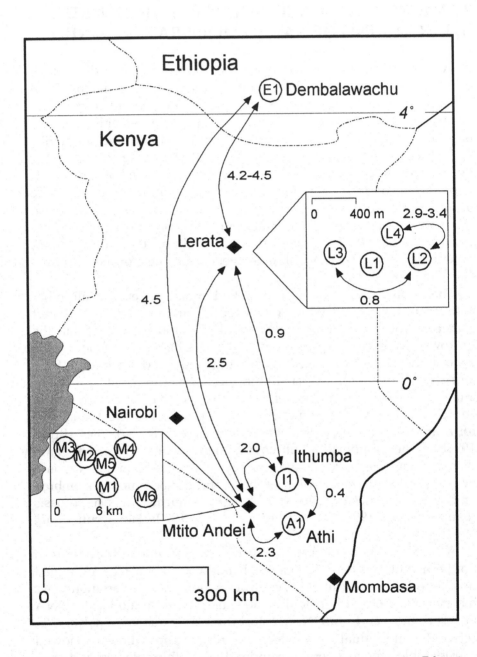

Figure 7.1. Map showing the relative locations of catching sites in Ethiopia (Dembalawachu), north Kenya (Lerata) and south Kenya (Mtito Andei, Athi and Ithumba) and the distribution of mitochondrial DNA haplotypes (labelled circles). Mean percent sequence differences (genetic distances) between haplotypes are indicated by arrows (adapted from Faukes *et al.*, 1997b).

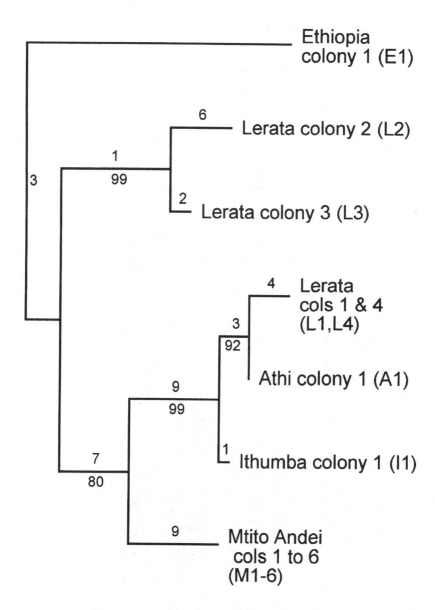

Figure 7.2. Phylogram showing relationships of the seven naked mole-rat haplotypes illustrated in Figure 7.1. The single tree was generated from maximum parsimony analysis of combined control region and cytochrome-*b* sequences, rooted with the Ethiopian haplotype as the outgroup, using the branch and bound search option in PAUP (Swofford, 1993). Numbers below each branch refer to the percentage bootstrap values following 100 replications, while values above are branch lengths (adapted from Faulkes *et al.*, 1997b).

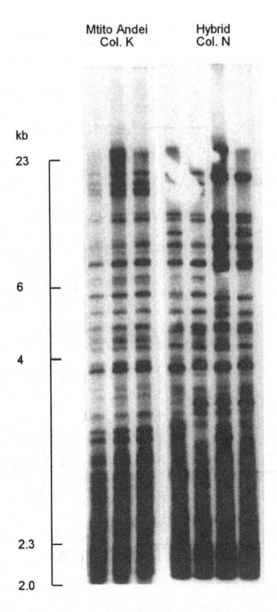

Figure 7.3. Multi-locus DNA fingerprints from minisatellite probe 33.6, following digestion with *Hin*fI, for seven individuals from two colonies. Colony K are the ancestors of a wild colony caught at Mtito Andei in Kenya (Colony M3 in Figure 7.1). Colony N are a hybrid colony formed from a cross between a northern and southern Kenyan animal. The increased levels of heterozygosity, reflected as reduced levels of band sharing, are readily apparent in colony N (data from Faulkes *et al.*, 1997b; C.G. Faulkes and M.W. Bruford, unpublished).

Individuals within colonies also had high mean coefficients of band sharing (the proportion of bands shared between individuals), estimated from DNA fingerprints, ranging from 0.93 to 0.99 (Faulkes *et al.*, 1997b). An example of this can be seen in Figure 7.3, showing three individuals from a laboratory colony 'K', which was directly descended from individuals caught in the wild in southern Kenya. These individuals were completely monomorphic (band sharing coefficient of 1.0) at the loci detected using this combination of enzyme digest and minisatellite probe (in this case *Hinf*I and 33.6 respectively). Normally, assuming Mendelian patterns of segregation and no linkage of minisatellite alleles, coefficients of band sharing among first degree relatives like parents–offspring or siblings would be approximately 0.5, while in monozygotic twins the value would be 1.0.

It could be argued that highly monomorphic fingerprints such as those of Colony K (Figure 7.3) might be produced as a result of linkage of minisatellite loci, rather than high levels of relatedness. However, family analysis of hybrid colonies of naked mole-rats, formed by crossing northern and southern Kenyan animals and containing increased levels of heterozygosity, confirms that linkage is not a significant factor. Patterns of inheritance of minisatellite loci between parents and offspring were found to be consistent with Mendel's Laws (Faulkes *et al.*, 1997b). DNA fingerprints for four individuals from one such hybrid colony 'N' are shown in Figure 7.3. The decrease in band sharing between individuals, compared with the 'wild-type' Colony K, is readily apparent. Bands common to individuals in both Colony K and Colony N represent the southern Kenyan alleles, while bands not present in Colony K represent the alleles contributed from the northern Kenyan stock.

In another study, Reeve *et al.* (1990) calculated from multi-locus DNA fingerprints that the average intra-colony relatedness (r) was 0.81 and the coefficient of inbreeding (F) was 0.45, the highest recorded for a natural mammalian population. It is argued that such close genetic relatedness between colony members would offset the individual reproductive sacrifice made by the non-breeders, most of which never breed, because their inclusive fitness is increased when the queen rears genetically similar offspring (Hamilton, 1964). Originally, the extreme inbreeding in naked mole-rats and subsequent high levels of genetic relatedness within colonies was thought to be a determining factor in the evolution of their eusocial behaviour (Reeve *et al.*, 1990; Faulkes *et al.*, 1990b, 1997b). Further studies of other bathyergid species have altered this view somewhat, and it is now thought that the high levels of inbreeding are a derived state (see Chapter 1) peculiar to this one species. All other bathyergids so far studied are obligate outbreeders and would be expected to have lower levels of relatedness. While the high levels of relatedness in naked mole-rats are an important factor to consider in our understanding of the evolution of eusociality in the Bathyergidae, it is apparent that ecological factors are also critically important (see Chapter 8).

While the occurrence of outbreeding events in wild colonies of naked mole-rats is rare due to the high risks involved in dispersal (including predation, energetic cost of burrowing and the xenophobic nature of colonies), limited outbreeding does appear to occur in the wild (Sherman *et al.*, 1992). The discovery of a male outbreeding phenotype in some captive colonies of naked mole-rats (O'Riain *et al.*, 1996) means that, in addition to the breeders, opportunities do exist for a small number of animals to increase their individual fitness. These are almost exclusively males who do not attempt to mate whilst within their natal colonies. In the wild, such dispersal events could also increase the genetic heterogeneity of colonies into which outbreeders migrate, potentially offsetting some of the deleterious effects of prolonged inbreeding. However, depending on the genetic composition of neighbouring colonies within a region, there could also be a fitness cost to non-breeders within the colony into which a disperser migrates. For example, a male disperser that is genetically different or less related to the new colony into which he migrates and reproduces would produce offspring that are in turn less related to the existing non-breeding helpers within that colony. The effect of this reduction in relatedness within the colony would be to reduce the inclusive fitness benefits gained by the non-breeding helpers and, assuming that kin recognition occurs, would potentially disrupt the social hierarchy within the colony. There is evidence to suggest that in captive colonies, queens may direct more aggression in the form of shoving behaviour at less related individuals (Reeve, 1992). Clearly, in order to understand the relative contributions of inbreeding, outbreeding and individual versus inclusive fitness in the social system of the naked mole-rat necessitates a knowledge of the genetic structure within and between colonies in the wild.

In two locations in Kenya, where multiple colonies have been studied, each had a different pattern of genetic structure. Six colonies from Mtito Andei in southern Kenya shared the same mitochondrial control region haplotype, suggesting a recent common maternal ancestor. In contrast, out of four colonies at Lerata in north Kenya, three haplotypes were identified. Two of these (Colony Lerata 2 versus Colonies Lerata 1 and 4) were quite divergent and phylogenetic analysis suggests that this area may be a zone where two distinct lineages are in close proximity (Figures 7.1 and 7.2).

These variable patterns of genetic structuring of colonies within geographic areas therefore have potentially different consequences for dispersal events. At Mtito Andei inter-colony genetic variability was low, with all six colonies sharing the same control region haplotype. The colonies were also found to have relatively high coefficients of band sharing in minisatellite DNA fingerprints. Low genetic variability among colonies at Mtito Andei has also been reported following RFLP analysis of mitochondrial DNA (Honeycutt *et al.*, 1991b). The occurrence of one haplotype in the colonies investigated implies that they are all derived from a common maternal ancestor distinct from that of the relatively close populations at Athi and Ithumba. The genetic similarity of the Mtito Andei colonies makes it difficult to determine, using these molecular techniques, whether or not any dispersal events have occurred. In the case of the Mtito Andei colonies, dispersal might be expected to be beneficial for both the disperser, who might gain an increase in direct fitness, and the recipient colony, which would gain a small increase in genetic variation. At the same time, the cost, in terms of loss in inclusive fitness to non-breeders within the recipient colony if the disperser breeds should be limited because these non-breeders would also share a high proportion of alleles in common with the disperser.

In contrast, at Lerata two colonies (Lerata 1 and 4) shared the same haplotype (Figure 7.1) which grouped in the same clade as the southern Kenyan colonies in the phylogenetic analysis (Figure 7.2). These two colonies were quite divergent from the other two haplotypes (found in Colonies Lerata 2 and 3, respectively) which only had a low level of divergence between them. These differences were also reflected in the coefficients of band sharing from the DNA fingerprints, despite the fact that these colonies were in close proximity (Faulkes *et al.*, 1997b). It is possible that these colonies are at the boundary of two distinct populations, one of which is derived from southern Kenya. These observations are also supported by separate allozyme studies of Lerata animals which revealed that one out of 34 loci showed heterozygosity among colonies. This is in contrast to Mtito Andei where individuals were homozygous for all 34 loci (Honeycutt *et al.*, 1991b). Furthermore, mtDNA RFLP analysis of individuals from Samburu, a region just southeast of Lerata, also showed increased inter-colony variation compared with Mtito Andei

(Honeycutt *et al.*, 1991b). Thus, the costs and benefits of dispersal at Lerata are potentially different from the Mtito Andei location. Because of the higher genetic heterogeneity, a greater fitness loss would be incurred among the non-breeding helpers of a colony accepting a male immigrant, because the subsequent offspring of the queen and the new male would be less related to the non-breeders. Unfortunately, from the data collected so far it is difficult to say which is the more typical pattern. More sampling of multiple colonies at different locations is now required to resolve this issue. However, the sister group relationships between the Mtito Andei, Ithumba, Athi and Lerata 1 and 4 haplotypes are suggestive of an isolation by distance model of divergence where 'budding' of colonies and limited dispersal and gene flow produced haplotypes that increase in sequence difference with increasing geographical distance. The extent to which naked mole-rats are able to discriminate one another on a genetic basis is also at present untested and until these questions are addressed the exact costs and benefits of dispersal, in terms of individual and inclusive fitness gains and losses, will not be clear.

As a result of both the tendency to avoid outbreeding and the subterranean niche of the naked mole-rat, gene flow must be limited and one would expect geographically distant populations to be highly divergent. This is supported by the data available, where Honeycutt *et al.* (1991b) used RFLP data to calculate an average nucleotide sequence divergence (*d*) of 5.4% between Samburu in the north of Kenya and Mtito Andei in the south. Results from both multi-locus minisatellite fingerprinting and sequence analysis of two mtDNA loci have shown that at a macro-geographic level colonies of naked mole-rats show considerable divergence, with fixation of different genotypes occurring at the different geographical locations over a wide proportion of their range (Figure 7.4; Faulkes *et al.*, 1997b).

DNA fingerprints and mtDNA haplotypes of samples collected from populations in Ethiopia, Somalia and Meru in northern Kenya were distinct from those from Lerata in northern Kenya and Mtito Andei, Athi and Ithumba in southern Kenya. An example of this can be seen in Figure 7.4, which again compares Colony K (from Mtito Andei in southern Kenya, Figure 7.1), with two mole-rats from a colony caught in Dembalawachu, Ethiopia and one mole-rat from a colony caught in Lerata, northern Kenya. There are no bands on the fingerprint in Figure 7.4 that can be unambiguously assigned as common to all locations. It is easy to see how the hybrid Colony N (Figure 7.3) has increased heterozygosity compared with Colony K, as Colony N was formed by crossing an animal from Mtito Andei

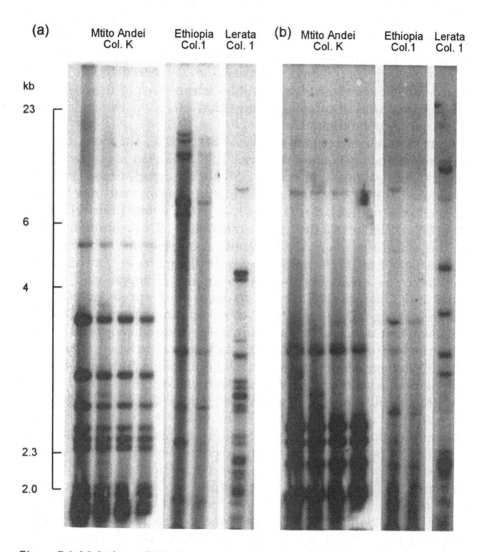

Figure 7.4. Multi-locus DNA fingerprints for seven individuals from three colonies from geographically distinct populations of naked mole-rats in Mtito Andei, southern Kenya, Lerata, northern Kenya, and Dembalawachu, Ethiopia. Samples were digested with restriction enzymes *Hinf*I and *Hae*III, and hybridised with minisatellite probe (a) 33.6 and (b) 33.15. Data adapted from Faulkes *et al.*, 1997b and C.G. Faulkes and M.W. Bruford, unpublished.

with one from Lerata. Both individuals have their own unique alleles, fixed in their respective populations, which segregate and recombine in their offspring to produce the observed genetic variability.

Similarly, most of the variance in mitochondrial DNA sequence divergence was found to be between geographic locations and there was a significant correlation between sequence divergence and geographic separation of haplotypes. The segregation of mtDNA haplotypes and minisatellite alleles within distinct areas probably reflects the high population viscosity (i.e. limited dispersal) and limited gene flow imposed by the subterranean niche of the mole-rat, and the patchiness of suitable habitats that can occur within an area (J.U.M. Jarvis, pers. comm.).

From the sequence data of Faulkes *et al.* (1997b), the mean genetic differences were maximum between the Ethiopian and southern Kenyan Mtito Andei haplotypes at 5.8% for the cytochrome-*b* gene. In vertebrates, mtDNA is cited as having an approximate divergence rate of 2% per million years (Moritz *et al.*, 1987), while Krajewski and King (1996) calculated a divergence rate of 0.7–1.7% per million years for the cytochrome-*b* gene of cranes. Assuming an approximate divergence rate of 2% per million years for the mole-rat cytochrome-*b* gene, then this would put the time since the Ethiopian and Mtito Andei individuals shared a common ancestor at approximately 2.9 million years ago. The fossil record of naked mole-rats in east Africa is known to extend back approximately three million years (Van Couvering, 1980), suggesting an ancient divergence between Ethiopian and Kenyan populations. The value of 5.8% divergence between Ethiopian and southern Kenyan haplotypes is slightly greater than the intra-specific range for cytochrome-*b* genetic distances known for other African mole-rats (Damaraland mole-rats: Namibian versus Botswana populations, approx. 700 km separation, 1.5%; common mole-rats: Cape Town versus Klawer populations in South Africa, approx. 230 km separation, 4.2%), and approaching that of some inter-specific distances (Cape dune mole-rat versus Namaqua dune mole-rat, 5.9%; Damaraland mole-rat versus *C. amatus*, 6.6%; Faulkes *et al.*, 1997a; Chapter 1). At the macrogeographic level, there was a clear trend for isolation by distance, with absolute geographic distance being significantly correlated with sequence divergence. As a consequence of this and other factors, such as long-term barriers to gene flow and limited female dispersal, widely dispersed naked mole-rat populations are highly divergent, although the available evidence suggests that *H. glaber* is still a monotypic species. Certainly, animals from Colonies Lerata 2 and 3 will readily breed with Mtito Andei individuals in captivity (e.g. Colony N in Figure 7.3) despite genetic distances of 3% between them at the cytochrome-*b* gene, although it is not known whether this would also hold true for

Ethiopian and Somalian individuals. The low species diversity of naked mole-rats is surprising given the high degree of genetic differentiation of conspecific populations when compared with other bathyergids, and the ancient ancestry of the naked mole-rat within the family (Allard and Honeycutt, 1992). In contrast, the sister genus *Cryptomys*, which inhabits a wider range of habitats, is relatively species rich, with at least 13 taxa with a variety of social structures (Faulkes *et al.*, 1997a; Chapter 1). While we have sampled individuals over a large proportion of the range of the naked mole-rat, populations in the far northern and eastern extremes in Ethiopia and Somalia are still to be investigated. Differences in the lengths of chromosomal short arms have been previously noted between Kenyan and Somalian individuals (Honeycutt *et al.*, 1991a).

In summary, the data for naked mole-rats suggest that fission of highly inbred colonies, coupled with rare male dispersal events, forms local populations that are genetically similar. Divergence increases with distance (e.g. Mtito Andei, Athi and Ithumba populations) or maintenance of highly divergent matrilines within a small area (e.g. Lerata). Ultimately, the limits to gene flow imposed by the subterranean niche have led to populations with sequence divergence approaching inter-specific levels.

7.3 INTRA-SPECIFIC GENETIC STUDIES OF OTHER BATHYERGIDS

After studies of relatedness in naked mole-rats, and the discovery of eusociality in Damaraland mole-rats, attention focused on determining the possible mating system and genetic relationships in the latter. Long-term mark–recapture studies in the wild (Jarvis and Bennett, 1993) and observations in captivity (Bennett, 1990) revealed some profound differences between these two species. While naked mole-rats appear to almost continually inbreed, the Damaraland mole-rat, and other cryptomids, have a strong inbreeding avoidance mechanism and will not mate with familiar conspecifics (Chapters 5 and 6).

Preliminary molecular genetic studies have confirmed that outbreeding is the norm in the Damaraland mole-rat. Six out of six breeding pairs that were examined were found to have distinct mitochondrial cytochrome-*b* haplotypes (Figures 7.5 and 7.6). The implication of this is that these breeding pairs are individuals from different maternal lineages and are therefore genetically divergent.

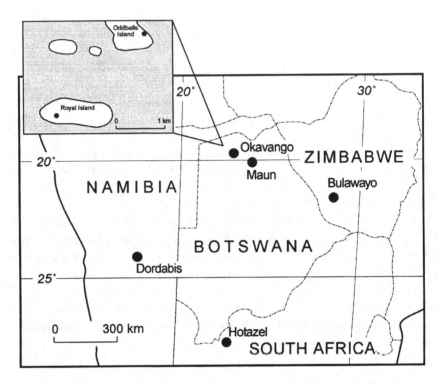

Figure 7.5. Map showing the relative locations of catching sites for the Damaraland mole-rats in Botswana, Zimbabwe, Namibia and South Africa used in genetic studies.

Table 7.1. Mean ± S.E.M. pairwise genetic distances of haplotypes within and between geographically distinct populations of the Damaraland mole-rat shown in Figure 7.5. Values are uncorrected 'p' distances calculated in PAUP (Swofford, 1993), n = number of pairwise comparisons (adapted from Faulkes *et al.*, 1997a; C.G. Faulkes, N.C. Bennett and J.U.M. Jarvis, unpublished).

	Botswana	Zimbabwe	Namibia	South Africa
Botswana	0.40 ± 0.06 $n = 15$	0.55 ± 0.06 $n = 6$	1.13 ± 0.04 $n = 24$	1.35 ± 0.05 $n = 24$
Zimbabwe	–	– $n = 1$	1.15 ± 0.10 $n = 4$	1.26 ± 0.10 $n = 4$
Namibia	–	–	0.44 ± 0.08 $n = 6$	0.88 ± 0.09 $n = 16$
South Africa	–	–	–	0.93 ± 0.19 $n = 6$

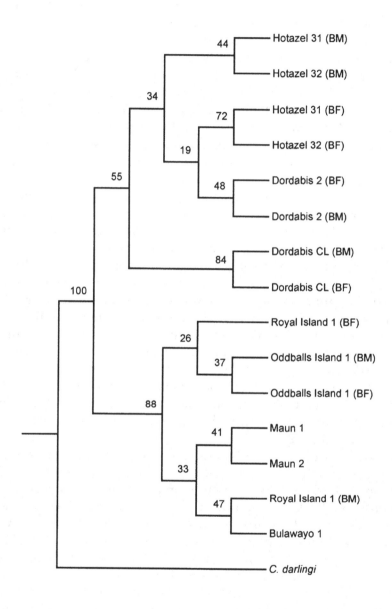

Figure 7.6. Cladogram showing relationships of 15 different Damaraland mole-rat haplotypes from nine colonies at locations illustrated in Figure 7.5. The tree was generated from maximum parsimony analysis of entire cytochrome-*b* sequences (1140bp), rooted with *C. darlingi* as an outgroup, using the dnapars option in PHYLIP (Felsenstein, 1989). Numbers above each branch refer to the percentage bootstrap values following 100 replications, BM and BF are breeding males and females respectively (data from Faulkes *et al.*, 1997a and C.G. Faulkes, unpublished).

Even very distant ancestors might be expected to share the same haplotype, given an average mutation rate of 2% per million years for mitochondrial DNA (Moritz *et al.*, 1987).

Compared with naked mole-rats, Damaraland mole-rats show much less sequence divergence between geographically separate populations (Table 7.1). For example, naked mole-rats from Lerata in northern Kenya and Dembalawatchu in Ethiopia, situated approximately 300 km apart, had levels of sequence divergence of between 4.2 and 4.5%, while between Mtito Andei in southern Kenya and Lerata, also around 300 km distant, there was 2.5% divergence (Figure 7.1). Damaraland mole-rats in the Okavango delta in Botswana and Bulawayo in Zimbabwe, separated again by around 300 km, had only 0.55% divergence at their cytochrome-*b* genes, while in animals from Botswana and Hotazel in South Africa, separated by about 1000 km, divergence was 1.35%. Although just looking at absolute geographic distances is somewhat simplistic, as geographical barriers such as rivers and mountains can effectively split neighbouring populations, levels of intra- and inter-population genetic divergence are consistently lower in the Damaraland mole-rat than in the naked mole-rat. This may reflect the more recent origin and radiation of the Damaraland mole-rat suggested by the molecular phylogeny (Figure 1.13), or perhaps a recent population bottleneck.

Despite the relatively low levels of sequence divergence, there were some clear trends in the phylogeographic structuring of haplotypes in the populations that were sampled. These relationships can be seen in the cladogram displayed in Figure 7.6. There was 100% bootstrap support for a split of the populations into two main sub-clades, one containing the South African and Namibian populations (from Hotazel and Dordabis respectively), and the second containing populations from the Okavango delta in Botswana and Bulawayo in Zimbabwe. These relationships are a good reflection of the relative geographical positions of these locations (Figure 7.5). Within the sub-clade containing animals from Botswana and Zimbabwe, there was also good bootstrap support (88%) for a further division (Figure 7.6) into a group containing the animals from the Oddballs Island colony, and one from Royal Island (the breeding female). The second group contained the breeding male from the Royal Island colony, the two animals from two colonies from Maun, and the individual from Botswana. This suggests that either the Royal Island breeding male is from a remnant population containing an ancestral haplotype, or that dispersal and gene flow have occurred.

The other sub-clade in Figure 7.6, containing the South African and

Namibian animals, showed haplotypes grouping in pairs according to their geographic origins. In the two Dordabis colonies, breeding pairs appear to have formed from individuals with similar haplotypes, and hence the breeding pairs in the two colonies ('Dordabis 2' and 'CL') each grouped together on the cladogram, although these two pairs of haplotypes were divergent from one another. On the other hand, at Hotazel, animals with the more divergent haplotypes formed breeding pairs in the two colonies studied. At both study sites, the colonies were all within a comparatively small area (500 m radius), and further studies are required to ascertain whether these trends are significant in any way.

Common mole-rats sampled in different populations in South Africa (Figure 7.7) showed different trends in their distribution of mitochondrial cytochrome-*b* haplotypes compared with the Damaraland mole-rat. Firstly, levels of sequence divergence were greater. In a study of eight animals from different locations in South Africa, where the entire cytochrome-*b* gene was sequenced, genetic distances ranged from 1.6% between individuals from Van Wyksdorp and Stellenbosch, to 4.5% between individuals from Somerset West and Stellenbosch. All of these locations were in relatively close proximity (Figure 7.7). A further screening of 696 base pairs of the 3' end of the cytochrome-*b* gene in 22 individuals taken from different colonies from locations shown in Figure 7.7 further illustrated the relatively high levels of both sequence divergence and haplotype diversity in the common mole-rat. Table 7.2 summarises the mean pairwise genetic distances for these individuals, while Figure 7.8 is a cladogram depicting their phylogentic relationships. Several interesting and informative points emerge from these results. For example, there is reasonable phylogeographic partitioning, with a major clade containing most of the Somerset West haplotypes, and a second containing most of those from Steinkopf. The single individuals from Klawer 1 and Steinkopf 7 both formed independent branches on the tree, the former having 100% bootstrap support for its separate grouping. Indeed, the genetic distances between this Klawer individual and the others were consistently higher, reaching a maximum of 5.8% between it and an animal from Somerset West 5. At present it is difficult to draw firm conclusions, based on one sample, as to why Klawer seems to be so distinct, and further investigation is required. If this individual is representative of the Klawer population as a whole, then it is possible that this region is divergent from other South African populations due to a founder effect or genetic drift.

Genetic structure of mole-rat populations

Despite the overall pattern, several haplotypes show deviation from a strict phylogeographic partitioning. For example, the Steinkopf 1 and Steinkopf 2 haplotypes are nested completely within a clade containing Somerset West haplotypes, and conversely, Somerset West 9 falls within a Steinkopf clade. The retention of ancestral haplotypes in different areas across the species distribution range seems the most likely explanation for the anomalous occurrence of these haplotypes. Alternatively, it could suggest that long-distance gene flow has occurred.

Table 7.2. Mean ± S.E.M. pairwise genetic distances of haplotypes within and between geographically distinct populations of the common mole-rat shown in Figure 7.7. Values are uncorrected 'p' distances calculated in PAUP (Swofford, 1993) from 696bp of the mitochondrial cytochrome-*b* gene, n = number of samples/number of pairwise comparisons (data from Faulkes *et al.*, 1997a; N. Oguge, C.G. Faulkes, N.C. Bennett, M.W. Bruford, J.U.M. Jarvis and A.C. Spinks, unpublished).

	Steinkopf $n=1$	Klawer $n=1$	Darling $n=1$	Somerset West $n=9$	Stellen-bosch $n=2$	Kuils-riviere $n=1$	Van Wyksdorp $n=1$
Steinkopf	2.56 ± 0.29 $n=21$	4.49 ± 0.02 $n=7$	3.33 ± 0.40 $n=7$	2.78 ± 0.17 $n=63$	2.34 ± 0.28 $n=14$	3.09 ± 0.45 $n=7$	2.00 ± 0.23 $n=7$
Klawer	–	–	5.42 $n=1$	4.74 ± 0.28 $n=9$	3.98 $n=2$	5.63 $n=1$	3.87 $n=1$
Darling	–	–	–	3.48 ± 0.11 $n=9$	4.48 $n=2$	3.33 $n=1$	3.79 $n=1$
Somerset West	–	–	–	2.39 ± 0.22 $n=36$	3.41 ± 0.28 $n=18$	3.63 ± 0.27 $n=9$	2.79 ± 0.26 $n=9$
Stellenbosch	–	–	–	–	1.5 $n=2$	3.89 $n=2$	2.5 $n=2$
Kuilsriviere	–	–	–	–	–	–	3.71 $n=1$
Van Wyksdorp	–	–	–	–	–	–	-

The most likely explanation for the observed patterns of haplotype distribution, diversity and divergence is a relatively high effective population size at both the local and meta-population level

(Figure 7.8; Table 7.2). Local effective population size in the common mole-rat would be expected to be higher than for the Damaraland mole-rat because there is greater turnover of colonies and increased chances of individual reproduction in this species (Spinks, 1998). Although reproductive skew is still high within colonies, i.e. only a single female and male breed, lower environmental constraints on dispersal are thought to result in a larger proportion of the population sucessfully breeding at some time in their life, compared with the Damaraland mole-rat. Thus, the effective population size is greater. Also, a higher turnover of reproductive animals means genes are less likely to become fixed in a population, and genetic diversity will be maintained. On a broader, meta-population level, effective population size may also be greater than for the Damaraland mole-rat, with the common mole-rat distributed in both mesic and arid habitats over a wider geographical range.

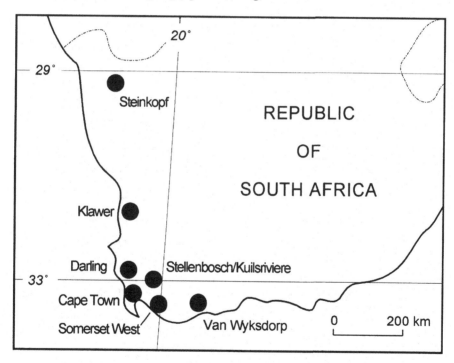

Figure 7.7. Map showing the relative locations of catching sites for the common mole-rats in South Africa used in genetic studies.

The limited population genetic data currently available therefore appear to reflect what we know of the social and reproductive

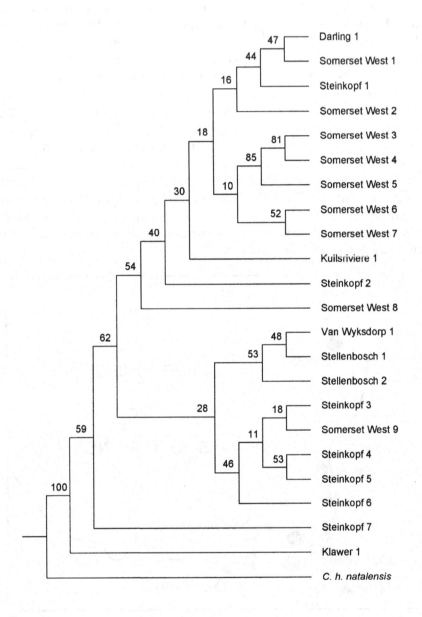

Figure 7.8. Cladogram showing relationships of the 22 different common mole-rat haplotypes each from different colonies/locations illustrated in Figure 7.7. The tree was generated from maximum parsimony analysis of cytochrome-*b* sequences, rooted with *C. h. natalensis* as an outgroup, using the dnapars option in PHYLIP (Felsenstein, 1989). Numbers above each branch refer to the percentage bootstrap values following 100 replications.

systems of African mole-rats. Inbreeding and high costs of dispersal in naked mole-rats produce colonies of animals that are almost totally homozygous, with local populations forming genetically distinct and divergent populations. Populations of common and Damaraland mole-rats, on the other hand, show levels of genetic diversity that would be expected from an outbreeding mammal. So far, results in the latter two species do not support the 'niche-width-genetic-variation' hypothesis of Nevo (1979), and observed patterns of genetic variation in all three species of mole-rats seem more explicable in terms of social structure, effective population size, and stochastic processes like genetic drift.

Chapter 8

The evolution of sociality in African mole-rats

8.1 WHAT IS A EUSOCIAL MAMMAL?

In the preceding chapters of this book we have discussed ecological, behavioural and physiological aspects of the social and reproductive systems of African mole-rats. In this final chapter we will examine these factors, together with our current understanding of the phylogenetic relationships within the family outlined in Chapter 1, and consider hypotheses for the evolution and maintenance of cooperative breeding and eusociality within the Bathyergidae. In order to tackle such questions, it is necessary to consider more generally the ways in which firstly families, then larger extended family groups of cooperatively breeding individuals, may evolve. It is also important to clarify some of the problems arising from semantics and definitions when attempting broad evolutionary comparisons of sociality, for example, when comparing different types of social vertebrates, as well as when comparing social vertebrates with social invertebrates. Most difficulties arise from attempting to classify and define specifically the level or degree of sociality, especially when extrapolating definitions first developed to describe insect societies. As mole-rats and most of the other cooperative breeding vertebrates considered in this book live in social groups where individuals are predominantly related to one another, we will be considering the evolution of eusociality from an inclusive fitness standpoint. Alternative theories have been proposed to explain altruism and cooperation in groups of unrelated individuals (for example, reciprocal altruism; Trivers, 1971) and the plethora of models that

211

followed based on game theory, like 'the prisoner's dilemma' and 'tit-for-tat' strategies. Detailed discussion of these is beyond the scope of this chapter, and the reader is referred to Dugatkin (1997) and Krebs and Davies (1997) for recent reviews of this field.

In Chapter 1, we introduced the original definition of eusociality that was derived from studies of insects, and to qualify, organisms were required to meet three criteria: a reproductive division of labour; overlap of two or more generations; and cooperative care of the young (Wilson, 1971). Naked mole-rats, *Heterocephalus glaber*, were the first mammal reported to fulfil these conditions (Jarvis, 1981), followed by Damaraland mole-rats, *Cryptomys damarensis* (Bennett and Jarvis, 1988b; Jarvis and Bennett, 1993). However, it could be argued that several other species should be included as eusocial vertebrates, such as the dwarf mongoose, *Helogale parvula* (Rasa, 1977; Rood, 1978). While they apparently fit the definition, they are generally not referred to as being eusocial. The same applies to species of cooperatively breeding birds like Florida scrub jays, *Aphelocoma coerulescens* (Woolfenden, 1975). There is clearly a problem, and it is made more complicated when trying to define 'grades' of sociality lower than that of eusocial. Michener (1969) attempted this for bees. He defined 'communal' as facultative groups of females of a single generation, sharing a common nest but each provisioning her own cells. 'Quasisocial' groups are small colonies in which two, or a small number, of females that may be of the same generation, cooperatively construct and provision cells. In 'semisocial' societies, small colonies show cooperative activity and there is a division of labour among adult females of the same generation, which distinguishes them from quasisocial groups. Michener also recognised that the ontogeny of a colony of a particular species may take it through more than one of these stages of sociality. This is an important point to bear in mind in social vertebrates as well, where changing environmental constraints may perhaps lead to changes in social structure. The spectrum of social organisation seen among insects is also often categorised into 'primitive' and 'advanced' eusociality (Wilson, 1971). Among cooperatively breeding vertebrates, the terms 'singular' and 'plural' have been suggested. In the former, reproduction is restricted to a single female or male, while in the latter, more than one male or female breeds. Some of the definitions and terminology adopted by sociobiologists working on cooperatively breeding and eusocial animals have been reviewed by Costa and Fitzgerald (1996) and Solomon and French (1997).

Given the myriad of definitions, can we really consider mole-rats or other vertebrates to be eusocial, or should the term be restricted to insect species to which the definition was first applied? Just how similar are 'eusocial' mole-rats to eusocial insects, that is, are they really extreme examples of convergent evolution, or just analagous to one another? These are questions that we will return to later in this chapter. Perhaps in the light of the discovery of highly social cooperatively breeding mammals, which were not well known when the social insects were first described, the definition of eusociality should be extended? There is certainly a school of thought that adheres to this premise. Crespi and Yanega (1995) suggest that the term 'eusocial' should be more rigorously applied to include only those species whose societies contain castes that are 'irreversibly behaviourally distinct'. This definition would exclude some invertebrates as well as all the social vertebrates, with, arguably, the naked mole-rats being an exception.

On the other hand, proponents of theories of optimal reproductive skew (introduced in Chapter 6) seek a more unified approach to the study of sociality that adopts a 'sliding scale' or continuum of sociality/eusociality (Sherman *et al.*, 1995). Thus, social groups are defined quite specifically by the amount of reproductive division of labour, providing a continuous variable useful in modelling and statistical analysis of social evolution. The index of reproductive skew is determined by calculating the variance in lifetime reproductive success among the members of a social group (Keller and Vargo, 1993; Reeve and Ratnieks, 1993). An argument against this approach is that only one aspect of sociality is quantified, namely the partitioning of reproduction. Crespi and Yanega (1995) suggest that lifetime reproductive success among individuals may not only be affected by social status, but also other factors such as higher mortality due to predation. However, it could also be argued that reproductive skew is itself influenced by a number of factors that include the level of constraints on dispersal, and relatedness, although the behavioural division of labour is not quantified in any way (Keller and Reeve, 1994). Keller and Perrin (1995) attempted to refine the reproductive skew approach to additionally quantify eusociality, and propose a 'eusociality index'. This is a measure complementary to reproductive skew, and attempts to quantify 'the degree to which some members of a society are specialized for reproduction, and others are specialized as helpers'. As with the reproductive skew index, values of the eusociality index vary from zero (non-eusocial) to one (eusocial). An alternative index of reproductive skew has also been presented by Pamilo and Crozier (1996), which attempts to eliminate some of the discontinuities that can occur using the equations of Keller and Reeve (1994). The problem with both reproductive skew and the eusociality index is that, at present, these parameters have actually been calculated in very few species (but see Keller and Perrin, 1995 and this chapter). In particular, lifetime reproductive success is notoriously difficult to ascertain. Recent advances in molecular genetic techniques, such as microsatellite genotyping, may help to redress this balance and provide powerful tools to assess parentage and reproductive skew without the need for behavioural observation. Such an approach has been applied to an investigation of the fine genetic structure of colonies of the ant, *Leptothorax acervorum*. By quantifying allelic variation at a number of microsatellite loci, Bourke *et al.* (1997) were able to determine parentage and reproductive skew in two out of eight multi-queened colonies that were studied.

It is clear from the preceding discussion that, despite the definition, the term eusocial means different things to different people, and the debate about how best to classify and quantify sociality seems set to continue. Strictly speaking, the criteria of Michener (1969) and Wilson (1971) could apply to many of the vertebrates with cooperative breeding systems. At the end of the day, categorical definitions such as those currently used to describe eusocial species are perhaps less useful for quantitative comparative studies than some kind of continuous measure of the type proposed by Sherman *et al.* (1995) and Keller and Perrin (1995).

8.2 THE EVOLUTIONARY ROUTES TO SOCIALITY

Theories on the evolution of eusociality in both invertebrates and vertebrates have been comprehensively reviewed by Alexander *et al.* (1991), with an emphasis on the origins of eusociality in the naked mole-rat. Since the publication of this manuscript in *The Biology of the Naked Mole-Rat*, a large body of new empirical and theoretical data has been accumulated from studies on mole-rats and other social vertebrates and invertebrates, which has shed further light on the evolution of cooperatively breeding strategies and eusociality. Setting aside social invertebrates, which often have rather specialised genetic mechanisms and life history strategies involved in shaping their societies, there are two principal ways in which societies may form. The parasocial route involves non-overlapping reproductive generations, or 'shared nests', composed of groups of related, or related and unrelated individuals. Michener (1969) first used the term parasocial to encompass his 'communal', 'quasisocial' and 'semisocial' groups of bees, although again some confusion arises as some 'semisocial' bees are also 'subsocial'. Vertebrate examples of parasocial societies include the banded mongoose, *Mungos mungo*, where communal suckling occurs among the pack members (Rood, 1974). Similarly, communal suckling and care of cubs in lion societies (Bertram, 1975) could also be included in this parasocial category.

The second principal way in which societes may form, and the most likely precursor of cooperatively breeding vertebrates, is the so-called subsocial route. In this case, groups of overlapping generations arise as a result of natal philopatry, where siblings delay dispersal and remain to form a family group. Thus, a family, as defined by Emlen (1995, 1997), is a social group where offspring continue to interact with their parents into adulthood, that is, beyond the age of sexual maturity. Emlen (1995, 1997) further differentiates families into simple or extended. In the latter, two or more group members, of either one or both sexes, breed. In simple families, only one pair breed, a situation akin to the social bathyergids, although with time immigration of non-family members into the group may also occur during dispersal (Jarvis and Bennett, 1993; Jarvis *et al.*, 1994; O'Riain *et al.*, 1996).

The choice of philopatry and family formation, or dispersal at weaning by a species has been investigated extensively and reviewed by Koenig *et al.* (1992) and Emlen (1994, 1995, 1997). This choice can be argued from the point of view of the costs of leaving and 'going it alone', or the benefits of staying, although the two are really inextricably linked to each other and should not be considered separately. From a Darwinian perspective, philopatry would be expected to evolve if the reproductive fitness benefits that accrue by staying at home exceed those of dispersal. If an individual disperses there is a risk involved in (i) finding a suitable territory due to habitat saturation, or perhaps the cost of establishing a territory (or burrow in the case of subterranean animals); (ii) obtaining a mate, and (iii) successful independent reproduction without helpers (Emlen, 1994; 1995). On the other hand, philopatry may increase the chances of survival for an individual, whilst providing a better chance of acquiring an opportunity to breed in a nearby territory, or inheriting a natal breeding position. Meanwhile, whilst in a family group, an individual will gain inclusive fitness benefits from aiding close relatives to breed successfully. The importance of some of these factors has been elegantly demonstrated by studies of cooperatively breeding birds. For example, family formation in acorn woodpeckers, *Melanerpes formicovorous*, has been clearly shown to be dependent upon natal territory quality, and the availability of nearby territories. The percentage of yearlings that delay dispersal declines rapidly as the proportion of territories becoming vacant increases (Emlen, 1984), while yearlings are more likely to delay dispersal if their natal territory quality is high (Stacey and Ligon, 1987). Similar trends have been reported in Seychelles warblers, *Acrocephalus seychellensis* (Komdeur, 1992), and a recent comparative analysis by Arnold and Owens (1998) suggests that in some birds, life history traits such as low annual mortality may predispose them to habitat saturation, and hence cooperative breeding.

For the subterranean mole-rats, the constraints of living underground add new dimensions to these choices that are not experienced by animals moving freely above ground. Many of these constraints are exacerbated by the effects of aridity in the habitats of the more social species like the Damaraland mole-rat and the naked mole-rat and, as we shall see, may play a central role in the evolution of eusociality in mole-rats. Habitat saturation may also play some role in limiting dispersal in mole-rats, but other ecological factors may have an overriding influence. While burrows may be densely distributed in some areas, and tunnels are known to come within one metre of each other in the case of Damaraland mole-rats, vacant sites adjacent to colonies that are numerically large are not always occupied straight away (Jarvis and Bennett, 1993; Jarvis *et al.*, 1998). Similarly, habitat saturation could potentially play a role enforcing natal philopatry in naked mole-rats, as colonies are known to occur in close proximity (Jarvis, 1985; Brett, 1991b; Faulkes *et al.*, 1997b). However, even if available sites are present, to exploit these, the conditions must be conducive to dispersal.

After natal philopatry has evolved in a species, there may be selection for the evolution of further traits that enhance reproductive fitness, usually indirectly, by the increased production and survival of close kin. Such traits may include offspring carrying in marmosets (Tardif, 1997), antipredator behaviour in dwarf mongoose (Rasa, 1986), and communal digging and foraging in mole-rats (e.g. the 'digging chains' of naked mole-rats) (Jarvis, 1981; Jarvis and Bennett, 1993).

Perhaps surprisingly, families are comparatively rare, and in a survey of the literature, Emlen found families in only 3% of mammal and bird species. However, some 96% and 90% of the familial birds and mammals, respectively, were found to have cooperative breeding stategies (Emlen, 1995, 1997). Amongst mammals, we have already mentioned the marmosets and tamarins of the family Callitrichidae, and dwarf mongooses and meerkats of the family Herpestidae. Cooperative breeding is also prevalent among canids, for example, the silver-backed jackal (Moelhman and Hofer, 1997) and wild dogs (Creel and Creel, 1995).

Within the Rodentia, cooperative or communal nesting and the care of young has been reported in 35 species from nine of the 30 families making up the order, although no phylogenetic trends are immediately apparent (see Solomon and Getz, 1997 for review). Social strategies in rodents range from plural breeders, which account for around 57% of cooperatively breeding rodents, to the more extreme singular breeders, with philopatry and reproductive suppression of non-breeders (remaining in the natal group for an extended period). There is, however, considerable inter- and intra-specific variation in these strategies. In singular breeding rodents, helping behaviour exhibited by non-breeders can be direct, such as huddling with young, and grooming or retrieving young straying from the nest, or indirect, such as obtaining food and building or maintaining nests and/or subterranean burrows. African mole-rats exhibit both of these forms of helping behaviour.

Among rodents, despite the relatively large number of species that exhibit cooperative breeding, apart from the Bathyergidae, few studies have examined group structure in detail on a long-term basis, or at least throughout the year. The prairie vole, *Microtus ochrogaster*, is one exception (Getz *et al.*, 1993). In addition, few studies have quantified either ecological variables or relatedness of individuals within groups. Singular breeding reaches its zenith in the Bathyergidae, where herbivory and a subterranean lifestyle in an expansive burrow system distinguishes the social bathyergid mole-rats from most, if not all, other cooperatively breeding vertebrates.

8.3 THEORIES OF SOCIAL EVOLUTION

The family structure and its patterns of relatedness are key features in the evolution and maintenance of cooperative breeding systems and eusociality. Kin selection, or selection favouring genes that promote altruistic (helping) behaviour towards individuals that are related to the altruist (Hamilton, 1963; Maynard-Smith, 1964), has long been thought to be a major selective force in the evolution of cooperative behaviour and reproductive altruism. In these situations, members of a non-breeding caste sacrifice their own reproduction in order to aid in the direct reproduction of others. The role of kin selection in insect sociality has been recently reviewed by Bourke (1997).

Hamilton's theory of inclusive fitness describes how genes that promote reproductive altruism and cooperation may be adaptive and spread within a population (Hamilton, 1964). Hamilton's ideas were a major turning point in the study of sociobiology, as the evolution of sterile castes had been a problem for biologists since Darwin, who initially thought that their occurrence may be 'fatal to my whole theory' (Darwin, 1859). Darwin subsequently went on to theorise how altruism may evolve, and his basic ideas and reasoning still stand today.

The central tenet of Hamilton's theory suggests that helping behaviour will be adaptive and undergo selection if a simple relationship is satisfied between the reproductive cost to an altruist in not producing its own offspring (c), the reproductive benefit to the recipient of the altruistic act (b), and the relatedness between the two individuals (r). If the conditions are such that $rb - c > 0$ (or $b/c > 1/r$), then helping or altruistic behaviour will be favoured (Hamilton, 1964; West-Eberhard, 1975). The key to understanding Hamilton's rule and subsequent theories on social evolution is that genes are shared between relatives simply as a result of inheritance according to Mendel's Laws. Therefore, the reproductive fitness of an individual, defined as a measure of the success of genes in perpetuating themselves in the course of evolution, will also increase if the individual helps relatives to reproduce successfully. This concept is known as inclusive fitness (Hamilton, 1964). In this scenario, selection could be said to be acting at the level of the gene, a classic example of 'the selfish gene' hypothesis, proposed by Dawkins (1976, 1989). So, while the costs of staying and remaining as a non-reproductive in a social group are reproductive suppression, and a reduction in personal fitness, this is offset by the inclusive fitness benefits that accrue when close relatives successfully breed. These 'intrinsic' or genetic factors have received much attention in the study of social evolution in invertebrates, while researchers on social vertebrates have had a tendency to focus on 'extrinsic' or ecological factors. However, the two are inextricably linked both in Hamilton's original theory, with extrinsic factors influencing the cost–benefit component, and in later theories of optimal reproductive skew (Keller and Reeve, 1994).

In Section 6.7 and at the beginning of this chapter, we introduced skew theory as a unified approach to studying the evolution of cooperative breeding and eusociality (Vehrencamp, 1983; Keller and Reeve, 1994; Sherman *et al.*, 1995). Skew theory considers cooperatively breeding animal societies as a continuum of social systems that differ in the way that lifetime reproductive success is distributed among group members, the so-called 'Eusociality Continuum' (Sherman *et al.*, 1995; Lacey and Sherman, 1997). In 'low skew' societies all individuals have an equal chance of reproduction, whereas in 'high skew' societies, reproduction is limited to one or a small number of individuals of each sex. Although all the social bathyergids have an extreme reproductive division of labour and are therefore by definition high skew societies, there are undoubtedly species differences in lifetime reproductive success. These appear to arise from differences in colony dynamics and dispersal opportunities, but remain to be quantified fully and systematically. We do know, however, that fewer naked mole-rats attain breeding status than Damaraland mole-rats (see page 184), and that common mole-rats may have an even greater chance of individual reproduction (Spinks, 1998).

Skew theory predicts that the reproductive skew and lifetime reproductive success in a society will be influenced by ecological constraints on independent breeding, the group's productivity if the subordinate stays within the group and cooperates, intra-group genetic relatedness, and the ability of the subordinate to fight and attain breeding status (Keller and Reeve, 1994). These predictions are generally supported by comparative studies on the various bathyergid species. In addition, skew theory assumes that dominant members of the society control reproduction amongst the subordinates, but that in certain circumstances it may pay the dominant to allow some subordinates to breed, as an inducement to stay in the group and cooperate peacefully. These 'staying incentives' are predicted to decrease as the ecological contraints increase, because the chances of successful independent breeding for subordinates dispersing are low. Secondly, inducements that prevent subordinates fighting for reproductive control may also occur, and these are known as 'peace incentives'. These would be expected when the fighting ability of subordinates is high and dominant–subordinate relatedness is low. We have seen in Chapters 4 and 5 that dominance relationships play a crucial role in the attainment of reproductive status in naked mole-rats (see also Clarke and Faulkes, 1997, 1998). Further reproductive conflicts of interest may also come into play in groups of mixed relatedness that arise from emigration and immigration, which are known to occur occasionally in both Damaraland mole-rats and naked mole-rats (J.U.M. Jarvis and N.C. Bennett, unpubl.; S. Braude, pers. comm.). To date it has been difficult to examine the relative contributions of the two kinds of reproductive inducement and test this aspect of skew theory in the Bathyergidae due to the confounding effect of incest avoidance. This would not apply to the naked mole-rat, but here the necessary field data are lacking. Thus, while skew theory attempts to provide a unifying principle for all cooperatively breeding societies, the assumptions of current models may not be totally universal. Incest avoidance is certainly an important factor to consider, and this alone may explain the reproductive skew observed in many societies. It is certainly an important factor in African mole-rats (see McRae *et al.*, 1997 and Clutton-Brock, 1998 for review of recent discussion on social evolution and skew theory).

Apart from theories with the central dogma that helping and cooperative breeding strategies are adaptive traits favoured by kin selection, other explanations for helper behaviour have been suggested that are non-adaptive. In these cases, care of young is suggested to occur as a result of an extension of parental behaviour. For example, in cooperatively breeding green woodhoopoes, *Phoeniculus purpureus*, Du Plessis (1993) has shown that adults responding to begging young will feed foreign nestlings that have been transferred to their nests as often as their own, and this is not influenced by the breeding status of the adult. The non-parental feeding behaviour was also carried out by recent-immigrant non-breeders, ruling out the possibility that feeding behaviour was related to kin discrimination, or a previous association with the breeders. Finally, helpers did not gain directly or indirectly any fitness benefits as a result of their feeding contributions. These results would seem to suggest that helping in this species is an unselected trait. Proponents of non-adaptive theories of helping also draw attention to the fact that evidence of helpers incurring costs as a result of their behaviour is generally lacking.

To test this argument, Clutton-Brock *et al.* (1998) used studies of wild meerkats, *Suricata suricatta*, in the Kalahari desert to investigate the potential costs of helping incurred by non-breeding meerkats during babysitting of young pups. Meerkats belong to the same family as the dwarf mongoose and also live in highly social groups of three to 20 individuals, with reproduction normally restricted to a dominant breeding female (Doolan and Macdonald, 1997). The young pups are cared for by other group members who act as 'babysitters', guarding the pups against predators and foreign conspecifics. During these bouts of babysitting, which normally last from dawn until dusk, babysitters do not feed and may lose 1% of their body weight, whereas non-babysitters may gain about 6% of their body weight over the same period (Clutton-Brock *et al.*, 1998). These losses may increase further if an individual continues babysitting for several days. Babysitters thus incur substantial costs as a result of their altruistic behaviour. Babysitters are normally the less dominant group members (breeding females rarely carry out babysitting), who are predominantly close relatives – 70% are offspring or siblings of the mother (Clutton-Brock *et al.*, 1998). Thus, in the case of meerkats, the low frequency of helping in dominant males and breeding females, together with the high cost of helping, strongly argues against a non-adaptive explanation for helping behaviour. The benefits of helping that offset these costs remain to be elucidated (Clutton-Brock *et al.*, 1998).

8.4 EVOLUTION OF SOCIALITY IN THE BATHYERGIDAE

Many studies on social behaviour have suggested that eusociality is an evolutionary endpoint that has arisen from a solitary ancestral form. Recent phylogenetic studies conducted upon insect taxa have revealed evidence of apparent evolutionary transitions in the opposite direction, namely, from eusocial to solitary behaviour. This has led to the idea that social evolution is bi-directional (Wcislo and Danforth, 1997).

We have seen from the phylogenetic data outlined in Chapter 1 (Figure 1.13) that the naked mole-rat is ancestral in the Bathyergidae, as its lineage was the first to branch off in the evolutionary tree of the family. This begs the question: what was the social status of the common ancestor of all bathyergids – was it solitary, colonial or eusocial? Jarvis and Bennett (1991) have suggested that the first bathyergids were probably solitary. This was because few, if any, other subterranean mammals are social (Nevo, 1979) and also because the earliest known fossil Bathyergoidea were large animals (Lavocat, 1978). The largest living bathyergids are solitary (Jarvis and Bennett, 1990) and by inference, this might suggest that, if these large fossil forms were completely subterranean, they were also solitary. However, H. Burda (pers. comm.) points out that there is no *a priori* reason to suppose the ancestor was solitary, as the social traits of a common ancestor could equally have been lost. He advocates a social ancestor, a trait he suggests was derived from their hystricognath ancestry. This so-called secondary solitarity has been reported in some species of bees (Wcislo and Danforth, 1997). If the common ancestor of bathyergids was solitary, then as Allard and Honeycutt (1992) first suggested, complex social behaviour has evolved twice in the family, once in the lineage leading to the naked mole-rat, and again in the common ancestor of the *Cryptomys* genus. On the other hand, if the common ancestor was social/eusocial, then this trait was retained in the naked mole-rats, and lost in the lineage leading to the radiations of the solitary mole-rats, and then evolved again in the common ancestor of *Cryptomys* (see Figure 1.13). In both scenarios complex social behaviour has independent origins within the family.

We cannot verify the status of the common ancestor directly, unless perhaps a fossilised colony of mole-rats, dated to the appropriate period, were to turn up. In the absence of such data, clues may be sought in the close hystricomorph relatives of the Bathyergidae (as suggested by Burda), or the 'outgroups' in the phylogeny. We have mentioned in Chapter 4 that there is evidence that some of the South American caviomorphs, like tuco-tucos (family Ctenomyidae), exhibit some form of social grouping. However, these New World hystricomorphs are divergent from the Old World families, and according to the molecular phylogeny of Nedbal *et al.* (1994), the closest relatives to the Bathyergidae are Old World porcupines (family Hystricidae), cane rats (family Thryonomyidae) and rock rats (family Petromuridae). There is evidence that some species in these families show signs of sociality, which could tip the balance in favour of a social common ancestor for the Bathyergidae. The molecular phylogeny of Nedbal *et al.* (1994) places Old World porcupines as ancestral to the Bathyergidae (see Figure 1.1). Interestingly, at least one species in this family, the Cape porcupine (*Hystrix africaeaustralis*) has colonial habits. Cape porcupines live in colonies of six to eight individuals, consisting of an adult pair and consecutive litters of offspring, which are normally singletons, occasionally twins or rarely triplets. Adult males protect the young, being aggressive towards foreign males and females, and accompany the young on foraging trips until 6–7 months of age, after which they tend to become solitary feeders (Van Aarde, 1987).

In the cane rats (family Thryonomyidae), both the greater cane rat (*Thryonomys swinderianus*) and the lesser cane rat (*T. gregarius*) are generally reported to be solitary, although individuals may live in close proximity in reed beds (Skinner and Smithers, 1990). Among the rock rats (family Petromuridae), *Petromus typicus* is reported to live in pairs or families in the crevices that occur in their rocky habitat, although information is somewhat vague (Skinner and Smithers, 1990). However, none of the species in these three families are subterranean, so it remains difficult to extrapolate from these and make any definite inferences about the ancestral bathyergid.

Whatever the nature of the ancestral bathyergid, the multiple occurrence of highly social behaviour makes them a powerful model system to examine the phylogenetic, genetic and ecological factors involved in the evolution of cooperative breeding strategies. Why are some species solitary, whereas others are social? Why does advanced sociality occur independently in two genera of the Bathyergidae? Why do the Bathyergidae differ from other subterranean rodents where a solitary existence is the norm (Nevo, 1979)? Several hypotheses have been proposed to explain the evolution of sociality in the African mole-rat family, and the principal contributing factors involved and their inter-relationships are summarised in Figure 8.1.

Prior to the discovery of eusociality in the outbreeding Damaraland mole-rat, inbreeding in naked mole-rats was thought to be an important determining factor in their eusocial behaviour (following the haplodiploidy genetic arguments developed from the studies of eusocial hymenopterans). With further knowledge of other mole-rats and cooperatively breeding vertebrates, it now appears that inbreeding is a derived trait peculiar to naked mole-rats. It is apparent that normal patterns of family relatedness, of 0.5 or less (first degree relatives), are sufficient for kin selection to operate and reproductive altruism to undergo selection, presumably if constraints on individual reproduction are sufficiently high. For example, average relatedness between subordinates and dominants in dwarf mongooses has been estimated as $r = 0.33$ (Creel and Waser, 1991; Keane *et al.*, 1994). To date, all the available evidence suggests that all other bathyergid mole-rats have inbreeding avoidance mechanisms and outbreed (Jarvis *et al.*, 1994; Burda, 1995; Bennett *et al.*, 1996b, 1997; Faulkes *et al.*, 1997a; Chapter 7). Thus, among cooperatively breeding vertebrates, the emphasis of research has now tended to shift away from looking at patterns of relatedness to examining ecological constraints. Among the bathyergids, both the costs of their subterranean lifestyle, and more proximate factors arising from their physiological and life history adaptations, form the foundations of theories of their social evolution.

Burda (1990) argues that relatively long periods of gestation,

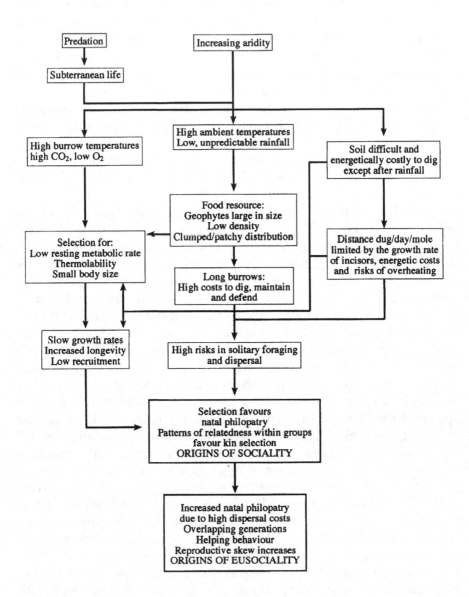

Figure 8.1. Summary of the principal factors, and their inter-relationships, thought to be important in the evolution of sociality and eusociality in the Bathyergidae. (modified from J.U.M. Jarvis, unpubl.; Brett, 1986 and Spinks, 1998).

altricial young and slow postnatal growth in the naked mole-rat and in '*Cryptomys hottentotus*' from Lusaka, Zambia (probably *C. amatus*), were phylogenetically determined, hystricomorph traits. Furthermore, Burda reports that female 'Cryptomys hottentotus' from Zambia cannot store fat reserves for gestation and lactational periods, and that given the lack of a subcutaneous fat layer in naked mole-rats (Jarvis, 1978), they too were unable to store fat. As a consequence, sociality would be adaptive, because females enduring a long pregnancy and lactation require provisioning of food from a cooperative workforce, on whom they would be totally dependent, as females from both species reduce their activity during gestation and lactation (Jarvis, 1969a; Burda, 1989). Further research has shown that the naked mole-rat does possess a thin subcutaneous fat layer (Buffenstein and Yahav, 1991a), and they are quite capable of storing fat, as can be seen in the disperser morph illustrated in Figure 5.7 (O'Riain et al., 1996). However, Urison and Buffenstein (1994) have shown that the energetic demands on the breeding queen during pregnancy and lactation are considerable, and that they do not appear to store appreciable quantities of energy. Body weight returns to non-pregnant levels immediately following parturition, and early pregnancy, potentially a period when energy could be stored, coincides with energy demanding lactation, due to the post-partum oestrous of naked mole-rats. While these findings support Burda (1990), similar data on fat/energy storage in the other bathyergid species are lacking. Furthermore, as we have seen in Chapter 5, there is no clear trend in increasing gestation length and sociality (Table 5.1), or post-natal development of pups and altriciality (Table 5.2). In addition, the observation that fat is not stored does not mean that the queen cannot, or that such an incapacity has led to sociality. Indeed, it could be argued that a low fat storage capacity in females could be a response to sociality, rather than a cause.

Alexander (1991) argues that group living arising from the subsocial route (parents and offspring) evolves as a result of predator avoidance, with parents protecting their brood from predation by placing them in a nest or protected environment like a burrow. This does seem a reasonable argument for the adoption of a subterranean lifestyle, but does not explain why natal philopatry and overlapping generations should necessarily follow, unless some other factor constrains dispersal, or there is increased protection against predators for large underground groups. If the latter is true, then an explanation must be sought for the fact that not all subterranean species are social, and why most lead a solitary existence, when the predators are often taxonomically similar for all subterranean rodents.

Recently, attention has focused on ecological factors that may correlate with sociality in mole-rats. Some general trends in the sociality and rainfall patterns of the habitat are immediately apparent. If the distributions of solitary, social and eusocial species of mole-rats are compared with mesic and arid vegetation types, the eusocial species are seen to occur in arid regions of unpredictable rainfall, while solitary genera are found in mesic parts of Africa where rainfall is higher and more predictable, and drought is rare (Figure 8.2; Jarvis *et al.*, 1994). These patterns of rainfall in turn give rise to characteristic patterns of geophyte distribution (see Chapter 3) and soil hardness, the latter then affecting the costs of burrowing. High energetic costs are therefore incurred as mole-rats extend their burrows both in search of food and in attempting to disperse and then find a mate and establish a new burrow system. These costs limit both of these activities to times when the costs of excavation and disposal of soil are at their lowest, that is, times when the soil has been moistened by rain and is easy to work (Vleck, 1979, 1981; Jarvis and Bennett, 1993; Jarvis *et al.*, 1998). It therefore follows that in arid regions, typically characterised by unpredictable rainfall and periods of drought, the opportunities open to the mole-rats for dispersal and for finding food are brief and widely spaced, whereas in mesic areas these opportunities are more numerous.

As we have discussed in Chapter 3, studies have also shown that, although the absolute amount of energy in the form of underground roots and tubers (geophytes) available to mole-rats is similar in arid and mesic areas (Bennett, 1988; Jarvis *et al.*, 1994; Table 8.1), the way in which it is distributed is different. Densities of geophytes are lower, but the absolute size of the food items is generally larger in the more arid regions than in the mesic ones. There is also evidence that the search for geophytes is blind and the mole-rats therefore cannot employ sensory cues to direct them to an energy-rich patch of food (Jarvis *et al.*, 1998; Chapter 3). Lovegrove and Wissel (1988) and Lovegrove (1991) developed a model to investigate the risks of unproductive foraging as a function of group size, and the density and size of the food resource. Their model predicted that cooperative foraging reduces the risks of unproductive foraging and thus represents a more stable long-term option in arid habitats where resources tend to be large but more widely dispersed. These factors would obviously also severely reduce the chances of successful dispersal by single mole-rats. Lovegrove and Wissel's (1988) model also suggested that the energetic benefits of sociality could only be realised if the total energy expenditure of the colony is minimised by

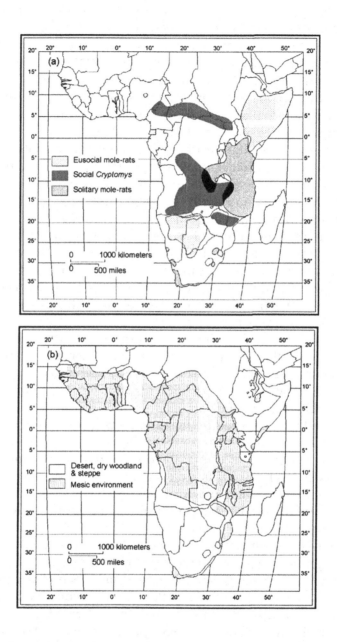

Figure 8.2. Approximate indication of (a) the distribution of solitary, social and eusocial mole-rats, and (b) rainfall patterns reflected as the distribution of arid and mesic vegetation types shown for sub-Saharan Africa only (adapted from Jarvis *et al.*, 1994).

reducing body size, mass-specific resting metabolic rates and thermoregulatory costs. They regarded the trend in the bathyergids towards mass-independent scaling of mass-specific resting metabolic rates to be an important factor in promoting sociality as a solution to environmental risk, and they termed it 'risk-sensitive metabolism' (see Chapter 3). Indeed, the thermoregulatory characteristics of the naked mole-rat, which resemble those of a poikilotherm, would be one example of this (Buffenstein and Yahav, 1991b).

For mole-rats inhabiting arid regions, both the limitation of few occasions when conditions are ideal for burrowing and the high risks that the solitary dispersers face in locating food would be expected to promote philopatry and group living. Alexander *et al.* (1991) have proposed that an important factor in the evolution of insect eusociality is the availability of a permanent nest, and they further suggest that the acquisition of a safe, expandable subterranean nest or burrow system, dug at considerable energetic cost to the colony, may have promoted sociality in subterranean rodents. This again would be a strong limitation to dispersal in arid regions. Even for social fossorial mammals such as pine voles, which do not have to dig to disperse, the building of a new burrow system is energetically expensive (Powell and Fried, 1992) and hence constitutes a significant constraint on successful dispersal. Other high risks would be incurred in locating a mate in these regions. As yet it is not known how this is achieved, but few animals are successful.

The advantages incurred by living in large groups in a harsh and unpredictable environment are well illustrated by ongoing field studies of a population of Damaraland mole-rats (Jarvis and Bennett 1993; Jarvis *et al.*, 1998). This study, currently in its tenth year, has coincided with a drought in which there have been years when periods of rainfall, and hence opportunities to disperse or locate new food resources, were spaced more than 10 months apart. Investigations have revealed that large colonies occupied territories that were significantly richer in resources than those of small colonies and that the failure rate of small colonies was much higher than that of large colonies (Jarvis *et al.*, 1998). Furthermore, large colonies continued to successfully raise pups throughout the periods of drought (Jarvis and Bennett, 1993; Jarvis *et al.*, 1998). Territory quality also affected the success of individuals in dispersing. Half as many individuals successfully dispersed, that is, were recaptured at least once after dispersing, from colonies in lower quality areas than from the higher quality areas (N.C. Bennett and J.U.M. Jarvis, unpubl.). This may be because individuals in the larger colonies (in higher quality territories) can afford to wait for the best opportunity to disperse. Cooperatively breeding acorn woodpeckers show a similar trend to these Damaraland mole-rats. The proportion of yearlings that delay dispersal increases markedly as natal territory quality increases from low to high (Stacey and Ligon, 1987).

8.5 ECOLOGICAL CONSTRAINTS AND SOCIAL EVOLUTION IN THE BATHYERGIDAE: COMPARATIVE ANALYSIS

More recently, there has been a synthesis of ideas based on the earlier models of Lovegrove and Wissel (1988) and Lovegrove (1991), on the risks of unsuccessful foraging and the observed trends in aridity and sociality. This current theory has become known as the food–aridity hypothesis, correlating sociality with habitat aridity and food distribution (Jarvis, 1985; Lovegrove and Wissel, 1988; Lovegrove, 1991; Jarvis *et al.*, 1994). As we mentioned earlier, solitary species tend to inhabit mesic areas, while the two eusocial species are found in arid regions with unpredictable rainfall (Figure 8.2; Table 8.1). The reduced rainfall generally leads to (i) harder soil and elevated costs of burrowing, and (ii) patchy food resources in the form of underground roots and the swollen tubers of arid-adapted plants, giving rise to an increased risk of unsuccessful foraging (Bennett, 1988; Lovegrove, 1991; Jarvis *et al.*, 1994). These high constraints on dispersal and individual reproduction are then hypothesised to lead to selection for group living, reproductive altruism, cooperative foraging and communal care of offspring.

When making broad species comparisons to test evolutionary hypotheses, as in the case of the mole-rats where we are comparing sociality with various ecological factors, individual species data points are often not independent. This is because a trait could have arisen simply by inheritance from a common ancestor, so-called phylogenetic bias, rather than by an adaptive response to selective pressures (Harvey and Pagel, 1991). So, for example, in the absence of phylogenetic knowledge, we might argue that the naked mole-rat and the Damaraland mole-rat share similar social and reproductive systems because they have shared common ancestry, rather than the trait evolving independently due to parallel evolution. Allard and Honeycutt (1992) first demonstrated that this was not the case and that these two species were quite divergent within the family.

Table 8.1. Summary of group sizes and ecological factors for the family Bathyergidae (Geophyte data as cited, or † (C.G. Faulkes, N.C. Bennett, J.U.M. Jarvis, A. Spinks and G.H. Aguilar, unpublished results); [1]Brett, 1991a,b; [2]Jarvis & Bennett, 1993; [3]1991; [4]Burda and Kawalika, 1993; [5]Bennett, 1988; [6]Lovegrove & Knight-Eloff, 1988; [7]Burda, 1990; [8]Bennett et al., 1994; [9]Bennett & Aguilar, 1995. Numbers in parentheses after the mean geophyte densities refer to the area (m²) that was surveyed. Mean annual rainfall values are grand means ± S.E.M. numbers in parentheses refer to the number of weather stations sampled for each species (reproduced from Faulkes et al., 1997a with permission from Proceedings: Biological Sciences).

Species	Max group size	Mean body mass (g)	Mean gestation (days)	Mean geophyte density ($n\,m^{-2}$)	Mean digestible energy ($kJ\,m^{-2}$)	Mean annual rainfall ($mm\,yr^{-1}$)	Mean no. months of >25 mm rainfall ($mon\,yr^{-1}$)	Rainfall coefficient of variation
Heterocephalus glaber	295[1]	34[3]	72[3]	0.08 ± 0.02 (3,600)[1]	204.5	362 ± 44 (9)	3.9 ± 0.3	79.2 ± 6.1
Cryptomys damarensis	41[2]	1064[3]	85[3]	3.7 ± 2.4 (9,300)[5,6,†]	179.1	390 ± 67 (6)	4.3 ± 0.6	59.2 ± 7.3
Cryptomys h. hottentotus	16[†]	77[3]	63[3]	278.8 ± 1.71(14)[†]	1342.7	538 ± 58 (3)	6.4 ± 1.0	33.5 ± 5.6
Cryptomys h. pretoriae	12[†]	–	–	–	–	704 ± 101 (1)	7.1 ± 0.9	27.0
Cryptomys amatus	10[†]	67[†]	100[7]	–	–	890 ± 39 (3)	5.6 ± 0.3	37.6 ± 2.8
Cryptomys darlingi	9[†]	65[†]	59[8]	37.2 ± 6.6 (10)[†]	1654.1	770 ± 49 (3)	5.8 ± 0.2	44.6 ± 2.0
Cryptomys mechowi	8[†,4]	233[†]	104[9]	173.0 ± 17.1 (20)[†]	2786.6	1129 ± 33 (4)	6.4 ± 0.3	44.5 ± 5.3
Cryptomys h. natalensis	8[†]	106[3]	–	–	–	872 ± 109 (3)	8.6 ± 0.5	47.1 ± 7.1
Cryptomys h. nimrodi	15[†]	–	–	–	–	–	–	–
Cryptomys 'choma'	4[†]	–	–	–	–	–	–	–
Cryptomys bocagei	4[†]	–	–	44.4 ± 20.1 (5)[†]	520.5	840 ± 20 (3)	6.4 ± 0.3	87.9 ± 38.7
Georychus capensis	1[3]	181[3]	46[3]	278.8 ± 1.71(14)[†]	1342.7	556 ± 47 (8)	7.6 ± 0.5	29.6 ± 2.3
Bathyergus suillus	1[3]	933[3]	52[3]	278.8 ± 1.71(14)[†]	1342.7	550 ± 71 (5)	7.5 ± 0.7	29.6 ± 3.5
Bathyergus janetta	1[3]	451[3]	–	18.8 ± 6.8 (318)[†]	376.8	82 ± 25 (3)	1.1 ± 0.5	23.5 ± 7.4
Heliophobius argenteocinereus	1[3]	160[3]	87[3]	–	–	913 ± 61 (4)	7.3 ± 0.5	81.8 ± 14.0

Using the phylogenetic trees constructed following mitochondrial DNA sequence analysis (see Chapter 1, Figure 1.13; Allard and Honeycutt, 1992; Faulkes *et al.*, 1997a), together with a number of ecological and life history factors (Table 8.1), trends in sociality within the Bathyergidae were analysed using CAIC (Comparative Analysis by Independent Contrasts; Purvis and Rambaut, 1995). The CAIC program enables us to quantitatively investigate the relationship between phylogeny, ecology and sociality within the Bathyergidae while controlling for phylogenetic bias. This analysis brought together many years of field and laboratory work, and has provided further support for the food–aridity hypothesis (Faulkes *et al.*, 1997a).

We have already mentioned the difficulties associated with defining the degree of sociality in cooperatively breeding species, and this problem remains with the Bathyergidae, despite the fairly detailed knowledge of group structure available for the various species. As a measure of sociality, the comparative study of Faulkes *et al.* (1997a) chose maximum group size, a parameter for which the most complete data set was available. Even so, group sizes for *Cryptomys* from Choma, Zambia, *C. bocagei* and *C. mechowi*, were based on limited data from three to six colonies. Maximum group size also encompasses a measure of natal philopatry, as the latter is a determinant factor in the former. The two species we currently define as eusocial also attain much larger maximum group sizes than the other species. Two Damaraland mole-rat colonies of over 40 animals have been captured to date (Jarvis and Bennett, 1993; J.U.M. Jarvis, N.C. Bennett and C.G. Faulkes, unpublished field data). Maximum group size in the naked mole-rat is even greater. One of the largest colonies found was caught by Rob Brett in 1984, near Kathekani in southern Kenya, and numbered at least 295 individuals (the breeding female was never caught) (Brett, 1991b). This particular colony was found in a cultivated field containing sweet potatoes, implying that abundant food may have been a factor in the production of this enormous group. However, food alone may not be the whole story, as one might predict that such rich pickings would lead to dispersal and new colony formation. Equally large colonies have also been captured in totally 'wild' naked mole-rat habitat at Meru, in northern Kenya. Between 1987 and 1996, Stan Braude caught 12 complete colonies containing over 120 animals, and of these, three had between 240 and 300 animals (S.H. Braude, pers. comm.).

Rather than maximum group size, the degree of reproductive skew, or the eusociality index described by Keller and Perrin (1995) would be a more appropriate measure of sociality to use in comparative analysis. Unfortunately, the data just are not currently available. Apart from the solitary species where we could assume skew to be zero (all have equal chances of reproduction), estimates of reproductive skew have only been made for two social species. In these cases high skew is associated with increased group size. The number of individuals within a colony that never attained breeding status were 99.9% for the naked mole-rat (n >4000 animals), and 92% for the Damaraland mole-rat (n >403 animals) (Jarvis *et al.*, 1994). Keller and Perrin (1995) have also estimated eusociality and reproductive skew indexes (E and S respectively) for a number of divergent species, including the naked mole-rat:

- Seychelles warbler $E = 0.02–0.76$ $S= 0.33–0.83$
- Dwarf mongoose $E = 0.5$ $S = 0.61$
- Halictid bee $E = 0.76–0.80$ $S= 0.93–1.00$
- Naked mole-rat $E = 0.96$ $S= 0.96$
- Honey bee $E = 1.00$ $S = 1.00$

As we can see, the naked mole-rat scores highly in both E and S, even in comparison to social insects. Measurement of reproductive skew, and the additional factors required for the energy/genes component of the eusociality index in all the bathyergids, is beyond our scope at present, but remains a challenge for the future.

Rainfall data for mole-rat habitats were obtained from the Global Historical Climatological Network, a database available via the Internet (www.ncdc.noaa.gov/ol/climate/research/ghcn/ghcn.html), and a total of 49 representative weather stations were chosen across the known ranges for all the mole-rat species that were included in the comparative analysis. Where possible these were close to field sites where colonies had been trapped. Great care must be taken when collating this type of data because in some regions rainfall can be very localised, and it is easy to be misled without detailed knowledge of the habitat. The monthly rainfall data available for each of these weather stations ranged over 22 to 124 years (with a mean of around 68 years of data). To give some idea of the 'useful' rainfall falling in the various habitats, the mean number of months per year having more than 25 mm of rain was ascertained, which gives at least an approximation of the time available for major burrowing activity. The figure of 25 mm is the estimated amount of rainfall needed to penetrate and soften the soil at the depth of most foraging tunnels (Jarvis *et al.*, 1994, 1998). To obtain a measure of the unpredictability of rainfall, a coefficient of variation was calculated by dividing the mean annual rainfall by its variance.

Food distribution parameters, proposed to correlate with sociality in the food–aridity hypothesis, were also included in the comparative analysis, and were obtained mainly from the literature, and from unpublished ecological surveys (N.C. Bennett, C.G. Faulkes, J.U.M. Jarvis and A. Spinks, unpublished results). Only geophytes known to be eaten by the mole-rats were quantified in the estimates of food distribution. Average figures for geophyte density were obtained either by extensive quadrating of an area inhabited by mole-rats (here the data were taken from a minimum of 20 to a maximum of 340 replicates of 0.5 m^2 to 1.0 m^2 quadrats), or complete surveying of an area containing colonies. The digestible energy available from these geophytes per unit area was calculated from the geophyte density values using published nutritional information for the different species of bulbs and tubers (Bennett and Jarvis, 1995). For the naked mole-rat, geophyte density values included all tubers eaten by them except *Macrotyloma* sp. For the latter, which form <5% of the total available biomass (Brett, 1991a), it was not possible to quantify their density due to their highly patchy distribution. Geophyte data used in the analysis for the common mole-rat, *Cryptomys h. hottentotus*, the Cape mole-rat, *Georychus capensis*, and the Cape dune mole-rat, *Bathyergus suillus*, were from an area where they occur sympatrically.

Rainfall and geophyte data used for the common mole-rat were for mesic habitats, although their range also extends into arid areas (the mean annual rainfall for three weather stations in such areas was 145.0 ± 39.3 mm yr^{-1}). However, the geophyte densities in these arid areas were relatively high at 78.8 ± 8.1 m^{-2} and social group sizes of two to 14 individuals are similar to mesic-adapted common mole-rat (A.C. Spinks, N.C. Bennett and J.U.M. Jarvis, unpublished results).

The data in Table 8.1, excluding gestation length, were used together with information describing the branching pattern of each of the three trees shown in Figure 8.3, to perform a series of three comparative analyses using CAIC. Each of these analyses produced seven phylogenetically independent standardised linear contrasts at the nodes designated A to G in Figure 8.3, between ecological factors and social group size (Table 8.2). These were then tested for statistical significance by simple regression through the origin, using the logarithm of maximum group size as the dependent variable (taking the logarithm helps to reduce the effect of outliers in the data). The phylograms displayed in Figures 8.3a and b are consensus trees, which represent the average of several trees qualitatively, and therefore have no associated branch length data quantifying the amount of difference between taxa. In these cases

(a)

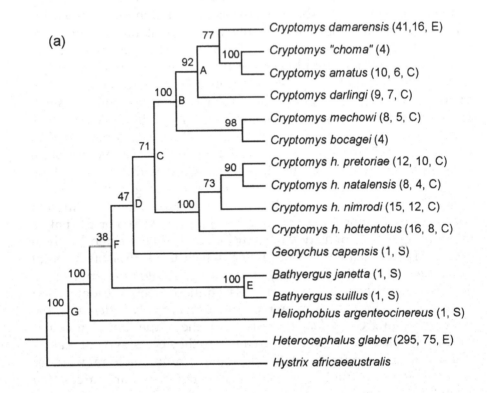

Figure 8.3. Phylogentic relationships of 15 African mole-rat haplotypes and one outgroup species (*H. africaeaustralis*) based on (a) parsimony analysis (PHYLIP) of combined 12S rRNA and cytochrome-*b* sequences. This is a consensus of two trees which differed only in the branching order of *Bathyergus* and *Heliophobius*, having a tree length of 2123 and consistency index of 0.57. Maximum social group sizes, and, where known, mean group sizes and social system (S = solitary, C = colonial with usually one breeding pair per colony, E = eusocial), are shown in parentheses; (b) consensus tree following parsimony analysis (PHYLIP) of 12S rRNA sequences, having a tree length of 662 and consistency index of 0.66. Diploid numbers (2n) are shown in parentheses (references as cited in Jarvis and Bennett, 1991; Aguilar 1993, unpublished; Macholán *et al.*, 1993), and (c) one of three trees generated by parsimony analysis (PAUP) of 12S rRNA sequences, having a tree length of 661 and consistency index of 0.66. Numbers above branches in (a) and (b) are % bootstrap values following 100 replications, while numbers in (c) are branch lengths. Letters A–G designate the seven internal nodes at which independent contrasts were calculated by CAIC.

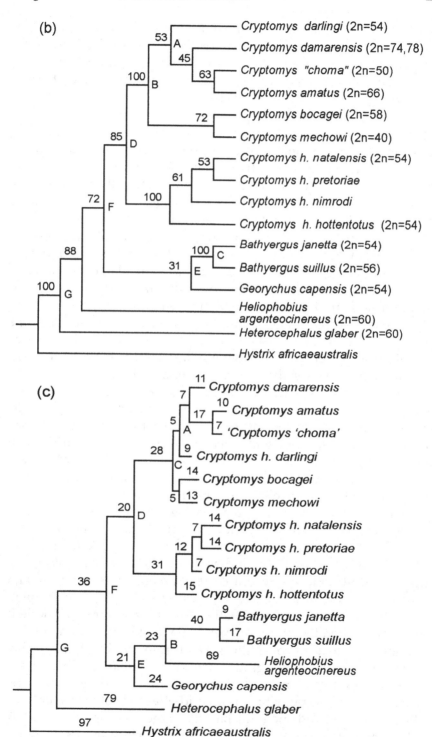

equal branch lengths were assumed in the analysis. However, in Figure 8.3c, the empirical branch length data obtained from the phylogenetic reconstruction were used in the comparative analysis, thus accounting for the amount of genetic difference between taxa as well as their phylogenetic relationships.

The comparative analyses produced results in support of the food–aridity hypothesis. In all three sets of analyses (Table 8.2; Faulkes *et al.*, 1997a), log maximum group size showed a negative correlation with log mean geophyte density (r = -0.778 to -0.822; p < 0.05), and was positively correlated with rainfall coefficient of variation (r = 0.882 to 0.915; p < 0.004). Digestible energy available per unit area did not correlate significantly with group size (p > 0.179). Thus, in the habitats that support the more social species, available energy in the form of geophytes is less densely distributed, a fact clearly shown in other studies that have calculated measures of geophyte clumping and patchiness for both naked and Damaraland mole-rats (Bennett, 1988; Lovegrove and Knight-Eloff, 1988; Brett, 1991a). While mean annual rainfall was lower in the habitats of the most social species, the naked and Damaraland mole-rats, there was no significant correlation. A confounding factor may have been inclusion of the data for the Namaqua dune mole-rat, *B. janetta*, an arid-adapted solitary species that inhabits extremely dry coastal regions, in apparent contradiction to the food–aridity hypothesis. However, in these areas the soil may be wetter than the rainfall data indicate, as coastal fogs and seepage areas provide additional unquantified precipitation and moistening of the soil (J.U.M. Jarvis, unpubl.). This may also explain why geophyte density and availible energy were high relative to the amount of rainfall in these areas.

A further analysis, using the tree topology in Figure 8.3a, included gestation length as an independent variable and produced six independent contrasts, as gestation length was not known for *B. janetta* (at nodes A to G, excluding E in Figure 8.3a; Table 8.2, 8.3a(ii)). In addition to significant correlations between log maximum group size and log mean geophyte density (p = 0.008), and rainfall coefficient of variation (p = 0.003), the loss of the contrast at node E (Figure 8.3) resulted in a third independent variable, the mean number of months per year of over 25 mm of rainfall, becoming highly negatively correlated with group size (p = 0.0009). This arises because the excluded node E included rainfall data from the Namaqua dune mole-rat, which is an arid-adapted solitary species.

Although the naked mole-rat has the largest group sizes and is also the smallest in body size, there were no clear trends in body

Table 8.2. Summary of regression statistics generated from independent standard linear contrasts between group size and ecological factors for the family Bathyergidae (values were calculated following CAIC analysis of the three phylogenetic trees displayed in Figure 8.3a, b and c). For analysis of the trees 8.3a and 8.3b equal branch lengths were assumed in the CAIC program (branch length data are not generated for consensus trees), while analysis of 8.3c used empirical branch lengths from parsimony analysis. The nodes A–G at which the independent contrasts were calculated by CAIC are also labelled on Figure 8.3a). Gestation length was included as an independent variable in 8.3a (ii) only, with the loss of one contrast at node E (Figure 8.3a). Statistics were obtained by regression (through the origin) of each parameter (independent variable) onto the log maximum group size (dependent) variable (reproduced from Faulkes *et al*, 1997a with permission from *Proceedings: Biological Sciences*).

Tree	Statistic	Log mean geophyte density	Log mean digestible energy	Log mean annual rainfall	Mean no. months of >25 mm rainfall	Rainfall coefficient of variation	Mean body mass	Gestation length
(i) 8.3a	r	-0.778	-0.527	0.003	-0.223	0.915	-0.469	–
	$F_{(1,6)}$	9.223	2.313	0.0001	0.315	31.050	1.690	–
	p	0.022*	0.179	0.993	0.595	0.001 ***	0.241	–
(ii) 8.3a	r	-0.884	-0.690	-0.476	-0.952	0.919	-0.576	0.522
	$F_{(1,5)}$	17.965	4.554	1.463	49.020	27.310	2.485	1.877
	p	0.008**	0.086	0.280	0.0009***	0.003**	0.175	0.229
8.3b	r	-0.822	-0.358	-0.092	-0.243	0.902	-0.328	–
	$F_{(1,6)}$	12.484	0.880	0.051	0.376	26.135	0.723	–
	p	0.012*	0.384	0.828	0.562	0.002**	0.428	–
8.3c	r	-0.787	-0.513	-0.190	-0.243	0.882	-0.201	–
	$F_{(1,6)}$	9.748	2.149	0.224	0.375	20.98	0.253	–
	p	0.021*	0.193	0.652	0.562	0.004**	0.633	–

mass (r = 0.201 to 0.576; p = 0.175 to 0.633) or gestation length (r = 0.522; p = 0.229) and sociality, as has been previously suggested for the Bathyergidae (Burda, 1990). Notwithstanding that energetic requirements, body size and gestational length may be important secondary/proximate factors that may influence behaviour (see Figure 8.1), it seems unlikely that they are ultimate determinants in social evolution.

In contrast to the other genera within the family, *Cryptomys* was relatively species rich with the 10 taxa examined in the comparative analysis exhibiting a variety of levels of sociality. Given that all the Bathyergidae are subject to high population viscosity and reproductive isolation due to the subterranean niche, one could ask the question, why has speciation not occurred to the same extent in the other genera? In the fire ant, *Solenopsis invicta*, it has been suggested that speciation may be driven by reproductive isolation and barriers to gene flow resulting from the development of alternative social organisations (Dewayne Shoemaker and Ross, 1996). While this hypothesis could explain the radiation of the social mole-rats, it may not explain species richness within *Cryptomys* (all taxa appear to be social, albeit to different degrees), and the lack of speciation in *Heterocephalus* (Faulkes *et al.*, 1997a). A possible explanation for the latter could be that the naked mole-rat is highly adapted to a very uniform and narrow ecological niche, whereas *Cryptomys* occurs over a much wider range of habitats, and with a predisposition to sociality in the common *Cryptomid* ancestor, they were able to undergo a rapid radiation into different niches.

The reduced genetic variation in natural populations of the naked mole-rat, a facultative inbreeder with only rare dispersal events (Reeve *et al.*, 1990; Faulkes *et al.*, 1990b, 1997b; O'Riain *et al.*, 1996), is also in complete contrast to the eusocial Damaraland mole-rat, where inbreeding is avoided and the incidence of dispersal and outbreeding is greater (Jarvis and Bennett, 1993; Jarvis *et al.*, 1994).

We are now able, for the first time, to consider phylogeny, ecology and relatedness together in analysing the causes of social evolution in the Bathyergidae, and show the importance of environmental constraints as a determining factor. It would seem that inbreeding is a derived trait peculiar to naked mole-rats, and studies in the Damaraland mole-rat suggest that, given a high enough level of ecological constraint, cooperative care of outbred siblings is sufficient for high skew, eusocial societies to evolve.

Further intra-specific comparative studies are now being undertaken to examine whether trends can be seen within the species inhabiting both mesic and arid habitats. For example, the common mole-rat is generally found in mesic areas, sometimes in sympatry with solitary mole-rats (Chapter 1). However, as we have mentioned, some populations are also found in very arid habitats, which begs the question, do these arid-adapted common mole-rats show any increased signs of sociality as a result of the environmental constraints? In a four-year study, Spinks (1998) found that while there were differences in food resources and their distribution between mesic and arid common mole-rat habitats, there were no clear differences in group sizes or reproductive characteristics. However, inter-population differences were apparent in both dispersal behaviour and xenophobic reactions between conspecifics. As one might predict from the food–aridity hypothesis, dispersal in common mole-rats in arid regions was found to be constrained and the temporal stability of colonies greater. These arid-dwelling common mole-rats were also markedly more aggressive to foreign conspecifics, compared with those in mesic areas. Again, this would be the prediction of a constraints model like the food–aridity hypothesis, because the fitness penalties incurred to non-breeders during an immigration event are greater in an arid habitat than in a mesic one, where opportunities for dispersal and individual reproduction are greater (Spinks, 1998). Similar investigations of Damaraland mole-rat populations in mesic areas of Zimbabwe could also be undertaken to provide further intra-specific comparisons.

It is perhaps surprising that convergent evolution has not produced highly social behaviour in any other subterranean rodents living in similar habitats to African mole-rats. It is interesting to speculate that perhaps this is because of the unique nature and distribution of the geophytes upon which mole-rats feed, together with climate and the underground niche, that is not found on any other continent.

8.6 EUSOCIALITY: VERTEBRATES VERSUS INVERTEBRATES

At the beginning of Chapter 1 and earlier in this chapter, we stated that two species of the Bathyergidae, the naked mole-rat and the Damaraland mole-rat, fulfil the basic criteria, derived from social insect studies, that define a eusocial society (Michener, 1969; Wilson, 1971). Indeed, as more data become available, other social cryptomids may fall within the eusociality fold. We have seen that ecological constraints and limited dispersal opportunities, and the patterns of relatedness within groups may all play a role. But how comparable are cooperatively breeding vertebrates, and specifically mole-rats, to eusocial invertebrates? Are vertebrate societies constrained by life history and allometric parameters, which mean they will never attain the size and degree of specialisation of invertebrates? Certainly, some of the characteristics of invertebrates and their societies are highly specialised and are not found among vertebrates. For example, the single mating flight and subsequent storage of sperm for months or years in queen bees means that the kin structure of colonies and the relatedness of individuals are predetermined and can be kept constant. In addition, the genetic asymmetries that arise from the haplodiploid system of sex determination are peculiar to the social Hymenoptera (bees, ants and termites), and may be complicated further, at least in bees, by the queen mating with more than one male.

It has been suggested in the past that genetic asymmetries that mimic haplodiploidy could occur in naked mole-rats as a result of alternate cycles of inbreeding (which we know occurs) and outbreeding (for which evidence exists). This genetic model was first proposed by Hamilton (1972) and further developed by Bartz (1979) as a possible explanation for the evolution of reproductive altruism in the diploid termites. Briefly, he proposed that if a highly inbred but unrelated 'king' and 'queen' termite were to reproduce, then their offspring would be heterozygous, but all for the same alleles. These individuals are therefore essentially genetically identical, and more closely related to each other than to each of the parents. As a consequence, these siblings can increase their inclusive fitness by increasing the production of brothers and sisters by the queen, in a similar way to haplodiploid invertebrates. Genetic studies of wild colonies of naked mole-rats which show that divergent populations may be in close proximity, for example, at Lerata, Kenya (Chapter 7; Faulkes *et al.*, 1997b), together with the known occurrence of outbreeding (S. Braude, pers. comm.; O'Riain *et al.*, 1996), support this idea.

While these kinds of genetic mechanisms cannot be ruled out as factors in naked mole-rat eusociality, in the other species of mole-rats and cooperatively breeding vertebrates, normal levels of relatedness (i.e. $r = 0.5$ or less) may be sufficient for reproductive altruism and eusociality to evolve, if the cost/benefit ratio of Hamilton's Rule is tipped in the right direction.

Colony size is one of the major differences between social

vertebrates and invertebrates and results from differences in life history variables like offspring production, rather than differences in sociality *per se*. Vertebrate societies are constrained to be small because of this. For example, colonies of African driver or 'army' ants, *Anomma wilverthi*, can contain up to a staggering 22 million workers, but possess just one breeding queen. These large colony sizes are achieved because the queen is able to produce some three to four million eggs in a month (Wilson, 1971).

In addition to the lower reproductive output of vertebrates, their lesser emphasis on 'hard-wired'patterns of behaviour, and the lack of the irreversible morphological and reproductive specialisation of invertebrates, social group size may also be constrained by factors relating to cognitive function, and the 'social glue' required to hold large groups of individuals together. In vertebrates, and especially mammals, where behavioural plasticity and social learning are characteristic, the individual requires a knowledge of 'who's who' around them, in order to make decisions on appropriate behavioural responses. In primates, some fascinating trends in relative brain size and social group size have been noted by Dunbar (1992). When the ratio of the size (volume) of the outer part of the brain, the neocortex, and the volume of the rest of the brain is plotted against the mean social group size for a given genus, the two were found to covary significantly. Extrapolating from this relationship, a group size in humans of 148 was predicted, a figure that is remarkably similar to the group sizes observed in traditional horticulturalist and hunter-gatherer societies (Dunbar, 1993). Among non-human primates, group cohesion is maintained through social grooming, and in Old World apes and monkeys, time spent grooming shows a linear relationship with group size (Dunbar, 1991). It is hypothesised that human societies can reach their relatively large sizes as a result of language, which enables numerous acts of 'social grooming' and bonding to occur simultaneously (Dunbar, 1993). The extent to which these kinds of relationships and arguments may apply in cooperatively breeding vertebrates, particularly non-primates, is unknown. In the case of naked mole-rats, their large group sizes result from selective pressures such as high constraints on dispersal, as we have discussed. Presumably, evolution of the necessary cognitive processes required to deal with these large group sizes must evolve in parallel, although research into this particular area is totally absent at present. Clearly, recognition of individuals and kin within the social group must be a crucial factor in the evolution and maintenance of social groups. The capacity for individual recognition impinges on many aspects of social behaviour, including kin selection, incest avoidance and nepotistic behaviours. A detailed discussion of kin recognition, and whether or not individual recognition cues (often odours) have a genetic or purely environmental basis (i.e. learnt familiarity) is beyond the scope of this chapter (see Pfennig and Sherman, 1995 for review). Among the Bathyergidae, research into this area is only just beginning to be undertaken and little is currently published. We have mentioned several times that naked mole rats are highly xenophobic to foreign conspecifics, even though they may be genetically similar and share common maternal ancestors - adjacent colonies in the wild may have arisen by the splitting and budding of an existing colony (Faulkes *et al.*, 1990a, 1997b; Reeve *et al.*, 1990; Brett, 1991a; Lacey and Sherman, 1997; Chapter 7). Passing behaviour in tunnels also appears to be very specific and related to dominance status. Captive studies have shown that on meeting face-to-face in a tunnel, naked mole-rats sniff one another's facial area and the dominant individual in the dyad then generally passes over the top of the subordinate (Clarke and Faulkes, 1997, 1998). The implication of these various observations is that naked mole-rats have mechanisms based on learnt recognition of individual and colony odours, and individual-specific vocalisations and vocalisations specific to encounters with foreign mole-rats, enabling them to recognise unfamiliar conspecifics (Pepper *et al.*, 1991; Judd and Sherman, 1996; O'Riain and Jarvis, 1997). The learnt recognition of familiar naked mole-rat colony members appears to be acute and may need frequent reinforcement by continual exposure of individuals to each other. In captivity it is often difficult to reintroduce a naked mole-rat that has been removed from its colony for periods of as little as 12 hours, as the absentee is treated as foreign and attacked (O'Riain and Jarvis, 1997).

Apart from individual recognition, communication is also a vital component for a society. We have mentioned that odour cues may play a role in individual recognition. In addition, the rich vocal repertoire of naked mole-rats may well be associated with their complex sociality. To date, 18 specific vocalisations have been characterised, more than for any other rodent species (Pepper *et al.*, 1991; Judd and Sherman, 1996). Among other social bathyergids, *Cryptomys* species from Lusaka, Zambia (probably *C. amatus*) are known to have at least 13 vocalisations, four of which were also recorded in a preliminary study of the giant Zambian mole-rat, *C. mechowi* (Credner *et al.*, 1997). Whether or not the solitary bathyergids have a more limited range of vocalisations remains unknown.

While the absolute sizes of social groups in vertebrates like mole-rats will never match those of some of the eusocial invertebrates, there is, nevertheless, a reasonable argument for considering that the upper end of the vertebrate social spectrum overlaps the bottom end of the invertebrate scale. Given the fundamental differences in the biology of these very divergent animal taxa, some common features do emerge. Some species of eusocial invertebrates show similarities with naked mole-rats, and vice versa. A good example of this is the reproductive conflict and dominance relationships seen in societies of the ponerine hunting ant in the *Diacamma* genus. These ants have no morphologically distinct queen but, instead, reproductive status in females is attained through dominance, as in the naked mole-rat. A more extreme strategy of suppression occurs in one of the Australian species, *D. australe*. All females are anatomically workers when they emerge from their cocoon, and bear bud-like vestigial wings called gemmae. It is the most dominant worker that lays the eggs, and she bites off the gemmae from her nest mates soon after they appear, which subsequently inhibits the development of their ovaries and thus permanently confines them to worker status. Reproduction is therefore restricted to a single dominant female, and only she mates with males (Peeters and Tsuji, 1993). In addition to this system of reproductive control by mutilation, dominant worker females of *Diacamma* species are also able to suppress ovarian activity of other females in the group, making them remarkably similar to naked mole-rats (Peeters and Tsuji, 1993). Colony sizes for *Diacamma* are also not dissimilar to those of naked mole-rats and other vertebrate societies (10 to 500 per colony).

A similar dominance-related monopoly of reproduction by breeding queens akin to that which we observe in naked mole-rats occurs in the facultatively eusocial stenogastrine hover wasps, *Liostenogaster flavolineata* (Field *et al.*, 1998). A colony foundress lays a few eggs and feeds the resulting larvae. Newly emerged adult females may then either disperse or become helpers in the natal nest. Group sizes are small with a population mean of one to four and, as with social vertebrates, there are no distinct morphological castes. In multi-female nests, one female is dominant and is thought to lay most, if not all, of the eggs, but helpers also stand a good chance of eventually becoming dominant and inheriting breeding status in the natal nest (Field *et al.*, 1998).

We have already mentioned, in Chapter 3, that naked mole-rats recruit colony mates to food sources, apparently by laying down an odour trail, in a way that bears an interesting similarity to eusocial bees dancing to communicate the whereabouts of food, and the odour trails laid down by some ant species (Judd and Sherman, 1996). Among naked mole-rats we can also see the beginning of the development of distinct morphotypes. The male disperser morph apparently has a particular body shape resulting from a distinctive distribution of body fat (Figure 5.7; O'Riain *et al.*, 1996). The body shape of the breeding queen is also quite distinct, with a characteristically elongated body which is often deeper dorso-ventrally as well (Jarvis *et al.*, 1991). These differences begin to develop during the growth spurt observed before and during the attainment of queen status. X-rays reveal the longer body of the queen is due to an increase in length of the vertebrae, which eventually become significantly longer than those of large non-breeding females (Jarvis *et al.*, 1991). These observations arguably qualify naked mole-rats as eusocial even by the more stringent criteria of castes that are 'irreversibly behaviourally distinct' as proposed by Crespi and Yanega (1995). Collectively, the large colony sizes and unusual morphological and behavioural traits of the naked mole-rat distinguish this species not only from the other bathyergids, but also from other cooperatively breeding vertebrates. Perhaps the naked mole-rat, more than any other species, deserves the reputation of an 'insect-like mammal'.

In summary, the Bathyergidae are an unparalleled mammalian group in which to study the evolution of sociality. The available evidence suggests that cooperative breeding in African mole-rats has been driven by two extrinsic environmental factors, these being the quantity and predictability of rainfall and hence aridity, and the subsequent density and clumping of their food resource. In mesic regions, where rainfall is fairly predictable and frequent, there is little restriction upon the time when burrowing can occur, and less risk of unsuccessful burrowing. In contrast, in arid regions, rainfall is erratic and unpredictable and consequently there is a limited time during which burrowing can effectively occur, and the reduced density of geophytes leads to an increased risk of unsuccessful foraging. Together, these factors produce high constraints on successful dispersal and hence individual reproduction, and selection for philopatry, altruistic behaviour and reproductive division of labour. Adaptation of intrinsic factors such as small body size and low metabolic rate may confer further selective advantages. Incest avoidance is apparently sufficient in most social mole-rat species to maintain a reproductive division of labour within colonies and ensure an adequate workforce. Obviously, colony efficiency would decrease considerably, even catastrophically, if all individuals were to reproduce.

The importance of an effective means of reproductive control, which is implicit in theories of optimal reproductive skew, is evident in the extreme physiological suppression mechanisms that have evolved in both male and female naked mole-rats, and in female Damaraland mole-rats. In naked mole-rats, because constraints on dispersal and therefore outbreeding are so high, inbreeding has become the norm, necessitating the strict reproductive control of the queen (see Chapter 6). Underlying the apparent harmony of naked mole-rat colonies is the strong motivation for individual reproduction which surfaces all too quickly during queen succession or take-overs, when inter-female aggression and conflict often lead to deaths. In other species, colonies undergo fission and dispersal occurs as soon as conditions are conducive. Thus the social bathyergid mole-rats are not the 'utopian societies' that they appear to be, and to quote Spinks (1998) are more realistically viewed as 'frustrated dispersers rather than content families'.

REFERENCES

Abbott, D.H. (1984) Behavioural and physiological suppression of fertility in subordinate marmoset monkeys. *Am. J. Primatol.* **6**, 169–186.

Abbott, D.H. (1987) Behaviourally mediated suppression of reproduction in female primates. *J. Zool. Lond.* **213**, 455–470.

Abbott, D.H., Hodges, J.K. and George, L.M. (1988) Social status controls LH secretion and ovulation in marmoset monkeys (*Callithrix jacchus*). *J. Endocrinol.* **117**, 329–339.

Abbott, D.H., Saltzman, W., Schutz-Darken, N.J. and Smith, T.E. (1997) Specific neuroendocrine mechanisms not involving generalized stress mediate social regulation of female reproduction in cooperatively breeding marmoset monkeys. *Ann. NY Acad. Sci.* **807**, 219–238.

Aguilar, G.H. (1993) The karyotype and taxonomic status of *Cryptomy hottentotus darlingi* (Rodentia: Bathyergidae). *S. Afr. J. Zool.* **28**, 201–204.

Alexander, R.D. (1974) The evolution of social behavior. *Annu. Rev. Ecol. Syst.* **5**, 325–383.

Alexander, R.D. (1991) Some unanswered questions about naked mole-rats. In *The Biology of the Naked Mole-Rat* (eds P.W. Sherman, J.U.M. Jarvis and R.D. Alexander), pp. 446–466. Princeton, NJ.

Alexander, R.D., Noonan, K.M. and Crespi, B.J. (1991) The evolution of eusociality. In *The Biology of the Naked Mole-Rat* (eds P.W. Sherman, J.U.M. Jarvis and R.D. Alexander), pp. 3–44. Princeton, NJ.

Allard, M.W. and Honeycutt, R.L. (1992) Nucleotide sequence variation in the mitochondrial 12S rRNA gene and the phylogeny of African mole-rats (Rodentia: Bathyergidae). *Mol. Biol. Evol.* **9**, 27–40.

Altuna, C.A. and Lessa, E.P. (1985) Penial morphology in Uruguayan species of *Ctenomys* (Rodentia, Octodontidae). *J. Mammal.* **66**, 483–488.

Altuna, C.A., Francescoli, G. and Izquierdo, G. (1991) Copulatory pattern of *Ctenomys pearsoni* (Rodentia, Octodontidae) from Balneario Solis, Uruguay. *Mammalia* **55**, 440–442.

Andersen, D.C. (1978) Observations on reproduction, growth and behaviour of the northern pocket gopher (*Thomomys talpoides*). *J. Mammal.* **59**, 418–422.

Andersen, D.C. (1982) Below ground herbivory: the adaptive geometry of geomyid burrows. *Am. Nat.* **119**, 18–28.

Andersen, D.C. (1987) Below ground herbivory in natural communities: a review emphasizing fossorial animals. *Quart. Rev. Biol.* **62**, 261–286.

Andersen, D.C. and MacMahon, J.A. (1981) Population dynamics and bioenergetics of a fossorial herbivore, *Thomomys talpoides* (Rodentia: Geomyidae) in a spruce fir sere. *Ecol. Monogr.* **51**, 179–202.

Andersson, M. (1981) Central place foraging in the whinchat, *Saxicola rubetra*. *Ecology* **63**, 538–544.

Anthony, A. and Foreman, D. (1951) Observations on the reproductive cycle of the black-tailed prairie dog (*Cynomys ludovicianus*). *Physiol. Zool.* **24**, 242–248.

Ar, A., Arieli, R. and Shkolnik, A. (1977) Blood-gas properties and function in the fossorial mole-rat under normal and hypoxic–hypercapnic atmospheric conditions. *Resp. Physiol.* **30**, 201–218.

247

Arieli, R. and Ar, A. (1981). Blood capillary density in heart and skeletal muscles of the fossorial mole-rat. *Physiol. Zool.* **54**, 22–27.

Arieli, R., Ar, A. and Shkolnik, A. (1977) Metabolic responses of a fossorial rodent (*Spalax ehrenbergi*) to simulated burrow conditions. *Physiol. Zool.* **50**, 61–75.

Arnold, K.E. and Owens, I.P.F (1998) Cooperative breeding in birds: a comparative test of the life history hypothesis. *Proc. R. Soc. Lond. B* **265**, 739–745.

Asa, C. S. (1997) Hormonal and experiential factors in the expression of social and parental behavior in canids. In *Cooperative Breeding in Mammals* (eds N.G. Solomon and J.A. French), pp. 129–149. Cambridge University Press, New York.

Aspey, W.P. (1977) Wolf spider sociobiology: I Agonistic display and dominance–subordinance relations in adult male *Schizocosa crassipes*. *Behavior* **62**, 103–141.

Avise, J.C. (1994) *Molecular Markers, Natural History and Evolution.* Chapman and Hall, New York.

Barnett, M. (1991) Foraging in the subterranean social mole-rat *Cryptomys damarensis*: a preliminary investigation into size-dependent geophyte utilisation and foraging patterns. Unpubl. Hons project, University of Cape Town, South Africa.

Bartz, S.H. (1979) Evolution of eusociality in termites. *Proc. Natl. Acad. Sci. USA* **76**, 5764–5768.

Batra, S.W.T. (1966) Nests and social behaviour of halictine bees of India. *J. Entomol.* **28**, 375–393.

Baudinette, R.V. (1972) Energy metabolism and evaporative water loss in the California ground squirrel. Effects of burrow temperature and water vapour pressure. *J. Comp. Physiol.* **81**: 57–72.

Bazli, G.O., Getz, L.L. and Hurley, S.S. (1977). Suppression of growth and reproduction of microtine rodents by social factors. *J. Mammal.* **58**, 583–591.

Bennett, N.C. (1988) The trend towards sociality in three species of southern African mole-rats (Bathyergidae): causes and consequences. Unpubl. PhD thesis, University of Cape Town, South Africa.

Bennett, N.C. (1989) The social structure and reproductive biology of the common mole-rat, *Cryptomys h. hottentotus* and remarks on the trends in reproduction and sociality in the family Bathyergidae. *J. Zool. Lond.* **219**, 45–59.

Bennett, N.C. (1990) Behaviour and social organization in a colony of the Damaraland mole-rat *Cryptomys damarensis*. *J. Zool. Lond.* **220**, 225–248.

Bennett, N.C. (1992) The social behaviour of a captive colony of the common mole-rat *Cryptomys hottentotus* from South Africa. *Z. Säugetierk.* **57**, 294–309.

Bennett, N.C. (1994) Reproductive suppression in eusocial *Cryptomys damarensis* colonies: a lifetime of socially-induced sterility in males and females. *J. Zool. Lond.* **234**, 25–39.

Bennett, N.C. and Jarvis, J.U.M. (1988a) The reproductive biology of the Cape mole-rat, *Georychus capensis* (Rodentia: Bathyergidae). *J. Zool. Lond.* **214**, 95–106.

Bennett, N.C. and Jarvis, J.U.M. (1988b) The social structure and reproductive biology of colonies of the mole-rat *Cryptomys damarensis* (Rodentia: Bathyergidae). *J. Mammal.* **69**, 293–302.

Bennett, N.C., Jarvis, J.U.M. and Davies, K.C. (1988) Daily and seasonal temperatures in the burrows of African rodent moles. *S. Afr. J. Zool.* **23**, 189–195.

Bennett, N.C., Jarvis, J.U.M. and Wallace, D.B. (1990) The relative age structure and body masses of complete wild-captured colonies of two social mole-rats, the common mole-rat, *Cryptomys hottentotus hottentotus* and the Damaraland mole-rat *Cryptomys damarensis*. *J. Zool. Lond.* **220**, 469–485.

Bennett, N.C., Jarvis, J.U.M, Aguilar, G.H. and McDaid, E.J. (1991) Growth rates and development in six species of African mole-rats (Rodentia: Bathyergidae) in southern Africa. *J. Zool. Lond.* **225**, 13–26.

Bennett, N.C., Clarke, B.C. and Jarvis, J.U.M. (1992) A comparison of metabolic acclimation in two species of social mole-rats (Rodentia: Bathyergidae) in southern Africa. *J. Arid Environ.* **22**, 189–198.

Bennett, N.C., Jarvis, J.U.M. and Cotterill, F.P.D. (1993a) Poikilothermic traits and thermoregulation in the Afrotropical social subterranean Mashona mole-rat, (*Cryptomys hottentotus darlingi*) (Rodentia: Bathyergidae). *J. Zool. Lond.* **231**, 179–186.

Bennett, N.C., Jarvis, J.U.M., Faulkes, C.G. and Millar, R.P. (1993b) LH responses to single doses of exogenous GnRH by freshly captured Damaraland mole-rats, *Cryptomys damarensis. J. Reprod. Fert.* **99**, 81–86.

Bennett, N.C., Jarvis, J.U.M. and Cotterill, F.P.D. (1994a) The colony structure and reproductive biology of the afrotropical Mashona mole-rat, *Cryptomys darlingi. J. Zool. Lond.* **234**, 477–487.

Bennett, N.C., Aguilar, G.H., Jarvis, J.U.M. and Faulkes, C.G. (1994b) Thermoregulation in three species of Afrotropical subterranean mole-rats (Rodentia: Bathyergidae) from Zambia and Angola and scaling within the genus *Cryptomys. Oecologia* **97**, 222–228.

Bennett, N.C., Jarvis, J.U.M., Millar, R.P., Sasano, H. and Ntshinga, K.V. (1994c) Reproductive suppression in eusocial *Cryptomys damarensis* colonies: socially-induced infertility in females. *J. Zool. Lond.* **234**, 617–630.

Bennett, N.C. and Aguilar, G.H. (1995) The colony structure and reproductive biology of the Giant Zambian mole-rat, *Cryptomys mechowi* (Rodentia: Bathyergidae). *S. Afr. J. Zool.* **30**, 1–4.

Bennett, N.C. and Jarvis, J.U.M. (1995) Coefficients of digestability and nutritional values of geophytes and tubers eaten by southern African mole-rats (Family, Bathyergidae). *J. Zool. Lond.* **236**, 189–198.

Bennett, N.C., Cotterill, F.P.D., Spinks, A.C. (1996a) Thermoregulation in two populations of the Matabeleland mole-rat (*Cryptomys hottentotus nimrodi*), and remarks on the general thermoregulatory trends within the genus *Cryptomys* (Rodentia: Bathyergidae). *J. Zool. Lond.* **239**, 17–27.

Bennett, N.C., Faulkes, C.G. and Molteno, A.J. (1996b) Reproductive suppression in subordinate, non-breeding female Damaraland mole-rats: two components to a lifetime of socially-induced infertility. *Proc. R. Soc. Lond. B* **263**, 1599–1603.

Bennett, N.C., Faulkes, C.G. and Spinks, A.J. (1997) LH responses to single doses of exogenous GnRH by social Mashona mole-rats: a continuum of socially-induced infertility in the family Bathyergidae. *Proc. R. Soc. Lond. B* **264**, 1001–1006.

Bennett, N.C. and Navarro, R. (1997) Differential growth patterns between successive litters of the Damaraland mole-rat, *Cryptomys damarensis* from Namibia. *J. Zool. Lond.* **241**, 185–202.

Bertram, B.C.R. (1975) Social factors influencing reproduction in wild lions. *J. Zool. Lond.* **177**, 463–482.

Beviss-Challinor, M. (1980) A preliminary investigation into the ecology of three species of sympatric mole-rats. Unpubl. project, Zoology Department, University of Cape Town, South Africa.

Blouin, S.F. and Blouin, M. (1988) Inbreeding avoidance behaviors. *Trends Ecol. Evol.* **3**, 230–233.

Bourke, A.F.G. (1988) Worker reproduction in higher eusocial Hymenoptera. *Quart. Rev. Biol.* **63**, 291–311.

Bourke, A.F.G. (1997) Sociality and kin selection in insects. In *Behavioural Ecology: an Evolutionary Approach*, 4th Edition, (eds J. Krebs and N. B. Davies), pp. 203–227. Blackwell Science Ltd, Oxford.

Bourke, A.F.G., Green, H.A.A. and Bruford, M.W (1997) Parentage, reproductive skew and queen turnover in a multiple-queen ant analysed with microsatellites. *Proc. R. Soc. Lond. B* **264**, 277–283.

Braude, S.T. (1991) The behavior and demographics of the naked mole-rat, *Heterocephalus glaber*. Unpubl. PhD thesis, University of Michigan, Ann Arbor. U.S.A.

Breed, W.G. (1976) Effect of environment on ovarian activity of wild hopping mice. *J. Reprod. Fert.* **47**, 395–397.

Brett, R.A. (1986) The ecology and behaviour of the naked mole-rat *Heterocephalus glaber* (Rüppell) (Rodentia: Bathyergidae). Unpubl. PhD thesis, University of London.

Brett, R.A. (1991a) The ecology of naked mole-rat colonies: burrowing, food and limiting factors. In *The Biology of the Naked Mole-Rat* (eds P.W. Sherman, J.U.M. Jarvis, and R.D. Alexander), pp. 137–184. Princeton University Press, Princeton, NJ.

Brett, R. A. (1991b) The population structure of naked mole-rat colonies. In *The Biology of the Naked Mole-Rat* (eds P.W. Sherman, J.U.M. Jarvis and R.D. Alexander), pp. 97–136. Princeton University Press, Princeton, NJ.

Broll, B.W. (1981) Comparative morphology of the gastro-intestinal tract of four species of mole-rat (Rodentia, Bathyergidae) in relation to diet. Unpubl. project, Zoology Department, University of Cape Town, South Africa.

Brown, J.L. (1987) *Helping and Communal Breeding in Birds*. Princeton University Press, Princeton, NJ.

Brown, R.E. and MacDonald, D.W. (1984) *Mammalian Social Odours*. Oxford University Press, Oxford.

Bruford, M.W. and Wayne, R. K. (1993) Microsatellites and their application to population genetic studies. *Curr. Opin. Genet. Dev.* **3**, 939–943.

Buffenstein, R. (1985) The effect of a high fibre diet on energy and water balance in two Namib desert rodents. *J. Comp Physiol.* **155**, 211–218.

Buffenstein, R. and Yahav, S. (1991a) Cholecalciferol has no effect on calcium and inorganic phosphorus balance in a naturally cholecalciferol-deplete subterranean mammal, the naked mole rat (*Heterocephalus glaber*). *J. Endocrinol.* **129**, 21–26.

Buffenstein, R. and Yahav, S. (1991b) Is the naked mole-rat *Heterocephalus glaber* an endothermic yet poikilothermic mammal? *J. Therm. Biol.* **16**, 227–232.

Buffenstein, R. and Yahav, S. (1994) Fibre utilization by Kalahari dwelling Damara mole-rats (*Cryptomys damarensis*) when fed their natural diet of gemsbok cucumber tubers (*Acanthosicyos naudinianus*). *Comp. Biochem. Physiol. A* **109**, 431–436.

Buffenstein, R., Laundy, M.T., Pitcher, T. and Pettifor, J.M. (1995) Vitamin-D-3 intoxication in naked mole-rats (*Heterocephalus glaber*) leads to hypercalcemia and increased calcium deposition in teeth with evidence of abnormal skin calcification. *Gen. Comp. Endocrinol.* **99**, 35–40.

Burda, H. (1989) Reproductive biology (behaviour, breeding and postnatal development) in subterranean mole-rats, *Cryptomys hottentotus* (Bathyergidae). *Z. Säugetierk.* **54**, 360–376.

Burda, H. (1990) Constraints of pregnancy and evolution of sociality in mole-rats. With special reference to reproductive and social patterns in *Cryptomys hottentotus* (Bathyergidae, Rodentia). *Z. Zool. Syst. Evolut.-forsch.* **28**, 26–39.

Burda, H. (1995) Individual recognition and incest avoidance in eusocial common mole-rats rather than reproductive suppression by parents. *Experientia* **51**, 411–413.

Burda, H. and Kawalika, M. (1993) Evolution of eusociality in the Bathyergidae: the case of the giant mole-rat (*Cryptomys mechowi*). *Naturwissenschaften* **80**, 235–237.

Burke T. and Bruford M.W. (1987) DNA fingerprinting in birds. *Nature* **327**, 149–152.

Busch, C., Malizia, A.I., Scaglia, O.A. and Reig, O.A. (1989) Spatial distribution and attributes of a population of *Ctenomys talarum* (Rodentia: Octodontidae). *J. Mammal.* **70**, 204–208.

Case, T.J. (1978) On the evolution and adaptive significance of postnatal growth rates in the terrestrial vertebrates. *Quart. Rev. Biol.* **53**, 243–282.

Catania, K.C. and Kaas, J.H. (1995) Organization of the somatosensory cortex of the star-nosed mole. *J. Comp. Neurol.* **351**, 549–567.

Charnov, E. (1976) Optimal foraging: the marginal value theorem. *Theor. Pop. Biol.* **9**, 129–136.

Clarke, F.M. and Faulkes, C.G. (1997) Dominance and queen succession in captive colonies of the eusocial naked mole-rat, *Heterocephalus glaber*. *Proc. R. Soc. Lond. B* **264**, 993–1000.

Clarke, F. M. and Faulkes, C.G. (1998) Hormonal and behavioural correlates of male dominance and reproductive status in captive colonies of the naked mole-rat, *Heterocephalus glaber*. *Proc. R. Soc.Lond. B* **265**, 1391–1399.

Clutton-Brock, T.H. (1988) Reproductive success. In *Reproductive Success* (ed. T.H. Clutton-Brock), pp. 472–485. Chicago University Press, Chicago.

Clutton-Brock, T.H. (1998) Reproductive skew, concessions and limited control. *Trends Ecol. Evol.* **13**, 288–292.

Clutton-Brock, T.H., Gaynor, D., Kansky, R., MacColl, A.D.C., McIlrath, G., Chadwick, P., Brotherton, P.N.M., O'Riain, J.M., Manser, M. and Skinner, J.D. (1998) Costs of cooperative behaviour in suricates (*Suricata suricatta*). *Proc. R. Soc. Lond. B* **265**, 185–190.

Cody, M.L., Breytenbach, G.J., Fox, B., Newsome, A.E., Quinn, R.D. and Siegfried, W.R. (1983) Animal communities: diversity, density and dynamics. In *Mineral Nutrients in Mediterranean Ecosystems* (ed. J.A. Day), pp. 91–110. South African National Scientific Programmes Report No. 71, June, 1983. CSIR, Pretoria, South Africa.

Cooper, H.M., Herbin, M. and Nevo, E. (1993) Ocular regression conceals adaptive progression of the visual system in a blind subterranean mammal. *Nature* **361**, 156–158.

Costa, J.T. and Fitzgerald, T.D. (1996) Developments in social terminology: semantic battles in a conceptual war. *Trends Ecol. Evol.* **11**, 285–289.

Cox, G.W. (1984) The distribution and origin of mima mound grasslands in San Diego County, California. *Ecology* **65**, 1397–1405.

Cracraft, J. (1983) Species concepts and speciation analysis. *Curr. Ornithol.* **1**, 159–187.

Credner, S., Burda, H. and Ludescher, F. (1997) Acoustic communication underground: vocalization charateristics in subterranean social mole-rats (*Crypromys* sp., Bathyergidae). *J. Comp. Physiol. A* **180**, 245–255.

Creel, S.R. and Waser, P.M. (1991) Failures of reproductive suppression in dwarf mongooses (*Helogale parvula*): accident or adaptation? *Behav. Ecol.* **2**, 7–15.

Creel, S.R., Creel, N., Wildt, D.E. and Monfort, S.L. (1992) Behavioral and endocrine mechanisms of reproductive suppression in Serengeti dwarf mongooses. *Anim. Behav.* **45**, 231–245.

Creel, S.R. and Creel, N.M. (1995) Communal hunting and pack size in African wild dogs *Lycaon pictus*. *Anim. Behav.* **50**, 1325–1339.

Creel, S.R. and Waser, P.M. (1997) Variation in reproductive suppression among dwarf mongooses: interplay between mechanisms and evolution. In *Cooperative Breeding in Mammals* (eds N.G. Solomon and J.A. French), pp. 150–170. Cambridge University Press, New York.

Crespi, B.J. and Yanega, D. (1995) The definition of eusociality. *Behav. Ecol.* **6**, 109–115.

Darden, T.R. (1972). Respiratory adaptations of a fossorial mammal, the pocket gopher, *Thomomys bottae*. *J. Comp. Physiol.* **78**, 121–137.

Darwin, C.R. 1859 (1967) *On the Origin of Species*: A facsimile of the First Edition with an Introduction by Ernst Mayr. Harvard University Press, Cambridge, Mass.

Davies, K.C. and Jarvis, J.U.M. (1986) The burrow systems and burrowing dynamics of the mole-rats *Bathyergus suillus* and *Cryptomys hottentotus* in the fynbos of the south-western Cape, South Africa. *J. Zool. Lond.* **209**, 125–147.

Davis-Walton, J. and Sherman, P.W. (1994) Sleep arrhythmia in the eusocial naked mole-rat. *Naturwissenschaften* **81**, 272–275.

Dawkins, R. (1976) *The Selfish Gene*. Oxford University Press, Oxford.

Dawkins, R. (1989) *The Selfish Gene*, 2nd Edition. Oxford University Press, Oxford.

De Graaff, G. (1964) A systematic revision of the Bathyergidae (Rodentia) of southern Africa. PhD thesis, University of Pretoria.

De Graaff, G. (1971) Family Bathyergidae. In *The Mammals of Africa: An Identification Manual* (eds J. Meester and H.W. Setzer), pp. 1–5. Smithsonian Institution Press, Washington DC.

De Graaff, G. (1972) On the mole-rat (*Cryptomys hottentotus damarensis*) Rodentia in the Kalahari Gemsbok National Park. *Koedoe* **15**, 25–35.

De Graaff, G. (1981) *The rodents of Southern Africa*. Butterworth, Johannesburg.

Dewsbury, D.A. (1975) Diversity and adaptation in rodent copulatory behavior. *Science* **190**, 947–954.

Dewsbury, D.A. (1982) Dominance, copulatory behaviour, and differential reproduction. *Quart. Rev. Biol.* **57**, 135–159.

DeWayne Shoemaker, D. and Ross, K.G. (1996) Effect of social organisation on gene flow in the fire ant *Solenopsis invicta*. *Nature* **383**, 613–616.

De Winton, W.E. (1896) On collections of rodents made by Mr. J. ffolliott Darling in Mashunaland and Mr. F.C. Selous in Matabeleland, with short field notes by the collectors. *Proc. Zool. Soc. Lond.* (52) 798–808.

De Winton, W.E. (1897) On a collection of rodents from Angola. *Ann. Mag. Nat. Hist.* pp. 320–324.

Dixon, K.W. (1981) Western Australian plants with underground fleshy storage organs. Unpubl. PhD thesis, University of Western Australia.

Doolan, S.P. and Macdonald, D.W. (1997) Band structure and failure of reproductive suppression in a cooperatively breeding carnivore, the slender-tailed meerkat (*Suricata suricatta*). *Behaviour* **134**, 827–848.

Dugatkin, L.A. (1997) *Cooperation Among Animals: An Evolutionary Perspective*. Oxford University Press, Oxford.

Dunbar, R.I.M. (1991) Functional significance of social grooming in primates. *Folia Primatol.* **57**, 121–131.

Dunbar, R.I.M. (1992) Neocortical size as a constraint on group size in primates. *J. Hum. Evol.* **22**, 407–421.

Dunbar, R.I.M. (1993) Coevolution of neocortical size, group size and language in humans. *Behav. Brain Sci.* **16**, 681–735.

Du Plessis, M.A. (1993) Helping behaviour in cooperative breeding green woodpeckers: selected or unselected trait? *Behaviour* **127**, 49–65.

Du Toit, J.T., Jarvis, J.U.M. and Louw, G.N. (1985) Nutrition and burrowing energetics of the Cape mole-rat, *Georychus capensis. Oecologia* **66**, 81–87.

Ellerman, J. R. (1940) *The families and genera of living rodents*, Vol 1. Trustees of the British Musuem (Natural History), London.

Eloff, G. (1958) The structural and functional degeneration of the eye of South African rodent moles *Cryptomys bigalkei* and *Bathyergus maritimus. S. Afr. J. Sci.* **54**, 293–302.

Emlen, S.T. (1984) Cooperative breeding in birds and mammals. In *Behavioral Ecology: An Evolutionary Approach*, 2nd Edition (eds J.R. Krebs and N.B. Davies), pp. 305–335. Blackwell Scientific Publications, Oxford.

Emlen, S.T. (1994) Benefits, constraints, and the evolution of the family. *Trends Ecol. Evol.* **9**, 282–285.

Emlen, S.T. (1995) An evolutionary theory of the family. *Proc. Natl. Acad. Sci.USA* **92**, 8092–8099.

Emlen, S.T. (1997) Predicting family dynamics in social vertebrates. In *Behavioural Ecology: An Evolutionary Approach*, 4th Edition (eds J.R. Krebs and N.B. Davies), pp. 228–253. Blackwell Science Ltd, Oxford.

Faulkes, C.G. (1990). Social suppression of reproduction in the naked mole-rat, *Heterocephalus glaber*. Unpubl. PhD thesis, University of London.

Faulkes, C.G., Abbott, D.H. and Jarvis, J.U.M. (1990a) Social suppression of ovarian cyclicity in captive and wild colonies of naked mole-rats, *Heterocephalus glaber. J. Reprod. Fert.* **88**, 559–568.

Faulkes, C. G., Abbott, D.H. and Mellor, A. (1990b) Investigation of genetic diversity in wild colonies of naked mole-rats by DNA fingerprinting. *J. Zool. Lond.* **221**, 87–97.

Faulkes, C.G. Abbott, D.H., Jarvis, J.U.M. and Sherriff, F.E. (1990c) LH responses of female naked mole-rats, *Heterocephalus glaber*, to single and multiple doses of exogenous GnRH. *J. Reprod. Fert.* **89**, 317–323.

Faulkes, C.G. and Abbott, D. H. (1991) Social control of reproduction in breeding and non-breeding male naked mole-rats (*Heterocephalus glaber*). *J. Reprod. Fert.* **93**, 427–435.

Faulkes, C.G., Abbott, D.H. and Jarvis, J.U.M. (1991) Social suppression of reproduction in male naked mole-rats, *Heterocephalus glaber. J. Reprod. Fert.* **91**, 593–604.

Faulkes, C.G. and Abbott, D.H. (1993) Evidence that primer pheromones do not cause social suppression of reproduction in male and female naked mole-rats (*Heterocephalus glaber*). *J. Reprod. Fert.* **99**, 225–230.

Faulkes, C.G., Trowell, S.N., Jarvis, J.U.M. and Bennett, N.C. (1994) Investigation of sperm numbers and motility in reproductively active and socially suppressed males of two eusocial African mole-rats, the naked mole-rat (*Heterocephalus glaber*), and the Damaraland mole-rat (*Cryptomys damarensis*). *J. Reprod. Fert.* **100**, 411–416.

Faulkes, C.G. and Abbott, D.H. (1997) Proximate mechanisms regulating a reproductive dictatorship: A single dominant female controls male and female reproduction in colonies of naked mole-rats. In *Cooperative Breeding in Mammals* (eds N.G. Solomon and J.A. French), pp. 302–334. Cambridge University Press, New York.

Faulkes, C.G., Bennett, N.C., Bruford, M.W., O'Brien, H.P., Aguilar, G.H. and Jarvis, J.U.M. (1997a) Ecological constraints drive social evolution in the African mole-rats. *Proc R. Soc. Lond. B* **264**, 1619–1627.

Faulkes, C.G., Abbott, D.H., O'Brien, H.P., Lau, L., Roy, M.R, Wayne, R.K. and Bruford, M.W. (1997b) Micro- and macro-geographic genetic structure of colonies of naked mole-rats, *Heterocephalus glaber*. *Mol. Ecol.* **6**, 615–628.

Felsenstein, J. (1989) PHYLIP – Phylogentic Inference Package, Version 3.2. *Cladistics* **5**, 164–166.

Field, J., Foster, W., Shreeves, G. and Sumner, S. (1998) Ecological constraints on independent nesting in facultatively eusocial hover wasps. *Proc. R. Soc. Lond. B* **265**, 973–977.

Filippucci, M.G., Burda, H., Nevo, E. and Kocka, J. (1994) Allozyme divergence and systematics of common mole-rats *(Cryptomys,* Bathyergidae, Rodentia) from Zambia. *Z. Säugetierk.* **59**, 42–51.

Filippucci, M.G., Kawalika, M., Macholan, M., Scharff, A. and Burda, H., (1997) Allozyme divergence and systematics of Zambian giant mole-rats, *Cryptomys mechowi* (Bathyergidae, Rodentia). *Z. Säugetierk.* **62**, 172–178.

Flynn, L.J. (1990) The natural history of Rhizomyid rodents. *Prog. Clin. Biol. Res.* **335**, 155–183.

Frank, L.G. (1986) Social organization of the spotted hyena *Crocuta crocuta*. II. Dominance and reproduction. *Anim. Behav.* **34**, 1510–1527.

French, J.A. (1997) Proximate regulation of singular breeding in Callitrichid primates. In *Cooperative Breeding in Mammals* (eds N.G. Solomon and J.A. French), pp. 34–75. Cambridge University Press, New York.

Gabathuler, U., Bennett, N.C. and Jarvis, J.U.M. (1996) The social structure and dominance hierarchy of the Mashona mole-rat, *Cryptomys darlingi* (Rodentia: Bathyergidae) from Zimbabwe. *J. Zool. Lond.* **240**, 221–231.

Gadagkar, R. (1994) Why the definition of eusociality is not helpful to understand its evolution and what should be done about it. *Oikos* **70**, 485–488.

Galef, B.G. and Buckley, L.L. (1996) Use of foraging trails by Norway rats. *Anim. Behav.* **51**, 765–771.

Galil, J. (1967) On the dispersal of the bulbs of *Oxalis cernua* Thunb. by mole-rats *(Spalax ehrenbergi)*. *J. Ecol.* **55**, 787–792.

Ganem, G. and Nevo, E. (1996) Ecophysiological constraints associated with aggression, the evolution towards pacifism in *Spalax ehrenbergi*. *Behav. Ecol. Sociobiol.* **38**, 245–252.

Gates, D.M. (1962) *Energy Exchange in the Biosphere*. Harper and Row, New York.

Gazit, I., Shanas, U. and Terkel, J. (1996). First successful breeding of the blind mole-rat *(Spalax ehrenbergi)* in captivity. *Isr. J. Zool.* **42**: 3–13.

Genelly, R.E. (1965) Ecology of the common mole-rat *(Cryptomys hottentotus)* in Rhodesia. *J. Mammal.* **46**, 647–665.

Gesser, H., Johanssen, K. and Maloiy, G.M.O. (1977) Tissue metabolism and enzyme activities in the rodent *Heterocephalus glaber*, a poor temperature regulator. *Comp. Biochem. Physiol.* **57**, 293–296.

Gettinger, R.D. (1984). A field study of activity patterns of *Thomomys bottae*. *J. Mammal.* **65**, 76–84.

Getz, L.L., McGuire, B., Pizzuto, T., Hofmann, J.E. and Fraise, B. (1993) Social organization of the pine vole *(Microtus ochrogaster)*. *J. Mammal.* **74**, 44–58.

Grant, W., French, N. and Folse, L. (1980) Effects of pocket gopher mounds on plant production in shortgrass prairie ecosystems. *Southwestern Nat.* **25**, 215–224.

Grodzinski, W. and Wunder, B.A. (1975) Ecological energetics of small mammals. In *Small Mammals: Their Productivity and Population Dynamics* (eds F.B. Golley, K. Petrusewicz and L. Ryszkowski). Cambridge University Press, Cambridge.

Hamilton, W.D. (1963) The evolution of altruistic behavior. *Am. Nat.* **97**, 354–356.

Hamilton, W.D. (1964) The genetical evolution of social behaviour. I, II. *J. Theor. Biol.* **7**, 1–52.

Hamilton, W.D. (1972) Altruism and related phenomena, mainly in social insects. *Annu. Rev. Ecol. Syst.* **3**, 193–232.

Hamilton, W.J. (1928) *Heterocephalus*, the remarkable burrowing rodent. *Brooklyn Mus. Sci. Bull.* **3**, 173–184.

Harrison, D. (1987) Preliminary thoughts on the incidence, structure and function of the mammalian vomeronasal organ. *Acta Oto. Laryngol.* **103**, 489–495.

Harrison, R.G. (1989) Animal mitochondrial DNA as a genetic marker in population and evolutionary biology. *Trends Ecol. Evol.* **4**, 6–11.

Harvey, P.H. and Pagel, M.D. (1991) *The Comparative Method in Evolutionary Biology.* Oxford University Press, Oxford.

Heffner, R.S. and Heffner, H. E. (1993) Degenerate hearing and sound localization in naked mole-rats (*Heterocephalus glaber*), with an overview of central auditory structures. *J. Comp. Neurol.* **331**, 418–433.

Heth, G. Frankenberg, E., Raz, A. and Nevo, E. (1987) Vibrational communication in subterranean mole-rats (*Spalax ehrenbergi*). *Behav. Ecol. Sociobiol.* **21**, 31–33.

Heuglin (1864) *Nov. Act. Ak. Caes. Leop.* Dresden XXXI.

Hickman, G.C. (1979) Burrow system structure of the bathyergid *Cryptomys hottentotus* in Natal, South Africa. *Z. Säugetierk.* **44**, 153–162.

Hickman, G.C. (1980). Locomotory activity of captive *Cryptomys hottentotus* (Mammalia: Bathyergidae), a fossorial rodent. *J. Zool. Lond.* **192**, 225–235.

Hickman, G.C. (1982) Copulation of *Cryptomys hottentotus* (Bathyergidae), a fossorial rodent. *Mammalia* **46**, 293–297.

Higgins, D. G., and Sharp, P.M. (1989) Fast and sensitive multiple sequence alignments on a microcomputer. *Cabios* **5**, 151–153.

Hill, W.C.O., Porter, A., Bloom, R.T., Seago, J. and Southwick, M.D. (1957) Field and laboratory studies on the naked mole-rat (*Heterocephalus glaber*). *Proc. Zool. Soc. Lond.* **128**, 455–513.

Hoelzel, A.R., Hancock, J.M. and Dover, G.A. (1991) Evolution of the cetacean mitochondrial D-loop region. *Mol. Biol. Evol.* **8**, 475–493.

Holldobler, B. and Wilson, E.O. (1990) *The Ants.* Cambridge, Massachusetts: Harvard University Press, Cambridge, Mass.

Honeycutt, R.L., Edwards, S.V., Nelson, K., and Nevo. E. (1987) Mitochondrial DNA variation and the phylogeny of African mole-rats (Rodentia: Bathyergidae). *Syst. Zool.* **36**, 280–292.

Honeycutt, R.L., Allard, M.W., Edwards, S.V. and Schlitter, D.A. (1991a) Systematics and evolution of the family Bathyergidae. In *The Biology of the Naked Mole-Rat* (eds P.W. Sherman, J.U.M. Jarvis and R.D. Alexander), pp. 45–65. Princeton University Press, Princeton, NJ.

Honeycutt, R.L., Nelson, K., Schlitter, D.A., and Sherman, P.W. (1991b) Genetic variation within and among populations of the naked mole-rat: Evidence from nuclear and mitochondrial genomes. In:*The Biology of the Naked Mole-rat* (eds P.W.

Sherman, J.U.M. Jarvis and R.D. Alexander), pp. 195–208. Princeton University Press,Princeton, NJ.

Howard, W.E. and Childs, H.E. (1959). Ecology of pocket gophers with emphasis on *Thomomys bottae mewa. Hilgardia* **29**, 277–358.

Hughes, R.N. (1979) Optimal diets under the energy maximisation premise: the effects of recognition time and learning. *Am. Nat.* **113**, 209–221.

Inouye, R., Huntley, N., Tilman, D. and Tester, J. (1987a) Gophers (*Geomys bursarius*), vegetation, and soil nitrogen along a successional sere in east central Minnesota. *Oecologia* **72**, 178–184.

Inouye, R., Huntley, N., Tilman, D., Stillwell, M. and Zinnel, K. (1987b) Old-field succession in a Minnesota sand plain. *Ecology* **69**, 12–21.

Irwin, D.M., Kocher, T.D., Wilson, A.C. (1991) Evolution of the cytochrome-*b* gene of mammals. *J. Molec. Evol.* **32**, 128–144.

Isil, S. (1983) A study of social behavior in laboratory colonies of the naked mole-rat (*Heterocephalus glaber* Ruppell; Rodentia, Bathyergidae). Unpubl. MSc thesis, University of Michigan, Ann Arbor.

Jacobs, D.S., Bennett, N.C., Jarvis, J.U.M. and Crowe, T.M. (1991) The colony structure and dominance hierarchy of the Damaraland mole-rat, *Cryptomys damarensis* (Rodentia: Bathyergidae) from Namibia. *J. Zool. Lond.* **224**, 553–576.

Jacobs, D.S. and Jarvis, J.U.M. (1996) No evidence for the work–conflict hypothesis in the eusocial naked mole-rat (*Heterocephalus glaber*). *Behav. Ecol. Sociobiol.* **39**, 401–409.

Jarvis, J.U.M. (1969a) Aspects of the biology of East African mole-rats. Unpubl. PhD thesis, University of East Africa, Nairobi.

Jarvis, J.U.M. (1969b) The breeding season and litter size of African mole-rats. *J. Reprod. Fert. Suppl.* **6**, 237–248.

Jarvis, J.U.M. (1973a) Activity patterns in the mole-rats *Tachyoryctes splendens* and *Heliophobius argenteocinereus. Zool. Afr.* **8**, 101–119.

Jarvis, J.U.M. (1973b) The structure of a population of mole-rats, *Tachyoryctes splendens* (Rodentia: Rhizomyidae). *J. Zool. Lond.* **171**, 1–14.

Jarvis, J.U.M. (1978) Energetics of survival in *Heterocephalus glaber* (Råppell), the naked mole-rat (Rodentia: Bathyergidae). *Bull. Carnegie Mus. Nat. Hist.* **6**, 81–87.

Jarvis, J.U.M. (1981) Eu-sociality in a mammal – cooperative breeding in naked mole-rat *Heterocephalus glaber* colonies. *Science* **212**, 571–573.

Jarvis, J.U.M. (1985) Ecological studies of *Heterocephalus glaber*, the naked mole-rat, in Kenya. *Natl. Geogr. Soc. Res. Rep.* **20**, 429–437.

Jarvis J.U.M. (1991) Reproduction of naked mole-rats. In *The Biology of the Naked Mole-Rat* (eds P.W. Sherman, J.U.M. Jarvis and R.D. Alexander), pp. 384–425. Princeton University Press, Princeton.

Jarvis, J.U.M. and Sale, J.B. (1971) Burrowing and burrow patterns of East African mole-rats *Tacyoryctes, Heliophobius* and *Heterocephalus. J. Zool. Lond.* **163**, 451–479.

Jarvis, J.U.M. and Bennett, N.C. (1990) The evolutionary history, population biology and social structure of African mole-rats: Family Bathyergidae. In *Evolution of Subterranean Mammals at the Organismal and Molecular Levels* (eds E. Nevo and O.A. Reig), pp. 97–128. Wiley Liss, New York.

Jarvis, J.U.M. and Bennett, N.C. (1991) Ecology and behavior of the Family Bathyergidae. In *The Biology of the Naked Mole-Rat* (eds P.W. Sherman, J.U.M. Jarvis and R.D. Alexander), pp. 66–96. Princeton University Press, Princeton, NJ.

Jarvis, J.U.M. and Bennett, N.C. (1993) Eusociality has evolved independently in two genera of bathyergid mole-rats – but occurs in no other subterranean mammal. *Behav. Ecol. Sociobiol.* **33**, 353–360.

Jarvis, J.U.M., O'Riain, M.J. and McDaid, E.J. (1991) Growth and factors affecting body size in naked mole-rats. In *The Biology of the Naked Mole-Rat.* (eds P.W. Sherman, J.U.M. Jarvis and R.D. Alexander), pp. 358–383. Princeton University Press, Princeton, NJ.

Jarvis, J.U.M., O'Riain, M.J., Bennett, N.C. and Sherman, P.W. (1994) Mammalian eusociality: a family affair. *Trends Ecol. Evol.* **9**, 47–51.

Jarvis, J.U.M., Bennett, N.C. and Spinks, A.C. (1998) Food availability and foraging by wild colonies of Damaraland mole-rats (*Cryptomys damarensis*): implications for sociality. *Oecologia* **113**, 290–298.

Jeffreys, A.J., Wilson, V. and Thein, S.L. (1985a) Hypervariable 'minisatellite' regions in human DNA. *Nature* **314**, 67–73.

Jeffreys, A.J., Wilson, V. and Thein, S.L. (1985b) Individual-specific 'fingerprints' of human DNA. *Nature* **316**, 76–79.

Johansen, K., Lykkeboe, G., Weber, R.E. and Maloiy, G.M.O. (1976) Blood respiratory properties in the naked mole-rat *Heterocephalus glaber*, a mammal of low body temperature. *Resp. Physiol.* **28**, 303–314.

Judd T.M. and Sherman, P.W. (1996) Naked mole-rats recruit colony mates to food sources. *Anim. Behav.* **52**, 957–969.

Kaufman, L.W. and Collier, G. (1981) The economics of seed handling. *Am. Nat.* **118**, 46–60.

Kayanja, F.I.B. and Jarvis, J.U.M. (1971) Histological observations on the ovary, oviduct and uterus of the naked mole-rat. *Z. Säugetierk.* **36**, 114–121.

Keane, B., Waser, P.M., Creel, S.R., Creel, N.M., Elliott, L.F. and Minchella, D.J. (1994) Subordinate reproduction in dwarf mongooses. *Anim. Behav.* **47**, 65–75.

Keller, L. and Vargo, E.L. (1993) Reproductive structure and reproductive roles in colonies of eusocial insects In *Queen Number and Sociality in Insects* (ed. L. Keller), pp. 16–44. Oxford University Press, Oxford.

Keller, L. and Reeve, H.K. (1994) Partitioning of reproduction in animal societies. *Trends Ecol. Evol.* **9**, 98–102.

Keller, L. and Perrin, N. (1995) Quantifying the level of eusociality. *Proc. R. Soc. Lond.* B **260**, 311–315.

Kennerly, T.E. (1964). Local differentiation in the pocket gopher (*Geomys personatus*) in southern Texas. *Texas J. Sci.* **6**, 297–329.

Kinloch, M.A. (1982) Behaviour and activity cycle of the common mole-rat, *Cryptomys hottentotus* Lesson 1826. Unpubl. project. Mammal Research Institute, University of Pretoria.

Koenig, W.D., Pitelka, F.A., Carmen, W.J., Mumme, R.L. and Stanback, M.T. (1992) The evolution of delayed dispersal in cooperative breeders. *Quart. Rev. Biol.* **67**, 111–150.

Komdeur, J. (1992) Importance of habitat saturation and territory quality for evolution of cooperative breeding in the Seychelles warbler. *Nature* **358**, 493–495.

Krajewski, C. and King, D. G. (1996) Molecular divergence and phylogeny: rates and patterns of Cytochrome *b* evolution in cranes. *Mol. Biol. Evol.* **13**, 21–30.

Krebs, J.R. (1978). Optimal foraging: decision rules for predators. In *Behavioural Ecology: An Evolutionary Approach*, First Edition (eds J.R. Krebs and N.B. Davies), pp. 23–63. Blackwell Scientific Press, Oxford.

Krebs, J.R. and Davies, N.B. (1991) *An Introduction to Behavioural Ecology.*, 3rd Edition. Blackwell, Oxford.

Krebs, J.R. and Davies, N.B. (1997) *Behavioural Ecology: An Evolutionary Approach.* Blackwell Scientific Publications, Oxford.

Kruczek, M. and Marchlewska-Koj, A.M. (1986) Puberty delay of bank vole females in a high-density population. *Biol. Reprod.* **35**, 537–541.

Kukuk, P.F. (1994) Replacing the terms 'primitive' and 'advanced'. New modifiers for the 'eusocial'. *Anim. Behav.* **47**, 1475–1478.

Kumar, S. and Hedges, S.B. (1998) A molecular timescale for vertebrate evolution. *Nature* **392**, 917–920.

Lacey, E.A., Alexander, R.D., Braude, S.H., Sherman, P.W. and Jarvis, J.U.M. (1991) An ethogram for the naked mole-rat: nonvocal behaviours. In *The Biology of the Naked Mole-Rat* (eds P.W. Sherman, J.U.M. Jarvis and R.D. Alexander), pp. 209–242. Princeton University Press, Princeton, NJ.

Lacey, E.A., and Sherman, P.W. (1991) Social organization of naked mole-rat colonies: Evidence for a division of labor. In *The Biology of the Naked Mole-Rat* (eds P.W. Sherman, J.U.M. Jarvis and R.D. Alexander), pp. 275–336. Princeton University Press, Princeton, NJ.

Lacey, E.A. and Sherman, P.W. (1997) Cooperative breeding in naked mole-rats: implications for vertebrate and invertebrate sociality. In *Cooperative Breeding in Mammals* (eds N.G. Solomon and J.A. French), pp. 267–301. Cambridge University Press, New York.

Lack, D. (1966) *Population Studies of Birds.* Clarendon Press, Oxford.

Lavocat, R. (1973) Les Rongeurs du Miocene d'Afrique Orientale. I Miocene Inferieur. *Mem. Trav. EPHE Institut de Montpellier* **1**, 1–284.

Lavocat, R. (1978) Rodentia and Lagomorpha. In *Evolution of African Mammals* (eds V.J. Magio and H.B.S. Cooke), pp. 69–89. Harvard University Press, Cambridge, Mass.

Leistner, O.A. (1967). The plant ecology of the southern Kalahari. *Botany Survey Memoir* **38**.

Lesson R.P. (1826) Voyage autour du monde sur la Coquille pendant 1822–1825. *Zoologie* **1**, 166.

Lovegrove, B.G. (1986a) Thermoregulation of the subterranean rodent genus *Bathyergus* (Bathyergidae). *S. Afr. J. Zool.* **21**, 283–288.

Lovegrove, B.G. (1986b) The metabolism of social subterranean rodents: adaptation to aridity. *Oecologia* **69**, 551–555.

Lovegrove, B.G. (1987). The energetics of sociality in mole-rats (Bathyergidae). PhD thesis, University of Cape Town.

Lovegrove, B.G. (1988) Colony size and structure, activity patterns and foraging behaviour of a colony of the social mole-rat *Cryptomys damarensis* (Bathyergidae). *J. Zool. Lond.* **216**, 391–402.

Lovegrove, B.G. (1989) The cost of burrowing by the social mole-rats (Bathyergidae) *Cryptomys damarensis* and *Heterocephalus glaber*: the role of soil moisture. *Physiol. Zool.* **62**, 449–469.

Lovegrove, B.G. (1991) The evolution of eusociality in mole-rats (Bathyergidae): a question of risks, numbers and costs. *Behav. Ecol. Sociobiol.* **28**, 37–45.

Lovegrove, B.G. and Jarvis, J.U.M. (1986) Coevolution between mole-rats (Bathyergidae) and a geophyte, *Micranthus* (Iridaceae). *Cimbebasia* **8**: 79–85.

Lovegrove, B.G. and Siegfried, W.R. (1986) Distribution and formation of mima-like earth mounds in the western Cape Province of South Africa. *S. African J. Sci.* **82**, 432–436.

Lovegrove, B.G. and Painting, S. (1987) Variations in the foraging behaviour and burrow structures of the Damara molerat *Cryptomys damarensis* in the Kalahari Gemsbok National Park. *Koedoe* **30**, 149–163.

Lovegrove, B.G. and Knight-Eloff, A. (1988) Soil and burrow temperatures, and the resource characteristics of the social mole-rat *Cryptomys damarensis* (Bathyergidae) in the Kalahari desert. *J. Zool. Lond.* **216**, 403–416.

Lovegrove, B.G. and Wissel, C. (1988) Sociality in mole-rats: metabolic scaling and the role of risk sensitivity. *Oecologia* **74**, 600–606.

Lovegrove, B.G., Heldmaier , G. and Ruf, T. (1993) Circadian activity rhythms in colonies of 'blind' mole-rats *Cryptomys damarensis* (Bathyergidae). *S. Afr. J. Zool.* **28**, 46–55.

Macdonald, D.W. (1984) *The Encyclopedia of Mammals.* Andromeda Oxford Ltd, Oxford.

Mácholan, M., Burda, H., Zima, A., Mísek, I. and Kawalika, M. (1993) Karyotype of the giant mole-rat, *Cryptomys mechowi* (Rodentia, Bathyergidae). *Cytogenet. Cell Genet.* **64**, 261–263.

McNab, B.K. (1966) The metabolism of fossorial rodents: a study of convergence. *Ecology* **60**, 1010–1021.

McNab, B.K. (1979) The influence of body size on the energetics and distribution of fossorial and burrowing animals. *Ecology* **60**, 1010–1021.

McRae, S.B., Chapuisat, M., and Komdeur, J. (1997) The ant and the lion: common principles and idiosyncratic differences in social evolution. *Trends Ecol. Evol.* **12**, 463–465.

Malizia, A.I. and Busch, C. (1991) Reproductive parameters and growth in the fossorial rodent *Ctenomys talarum* (Rodentia: Octodontidae). *Mammalia* **55**, 293–305.

Malizia, A.I and Busch, C. (1997). Breeding biology of the fossorial rodent *Ctenomys talarum* (Rodentia: Octodontidae). *J. Zool. Lond.* **242**, 463–471.

Matschie (1900) Eine neue abart von *Georychus* aus Togo, Deutsch-West-Africa. *Sitz.-Ber. Ges. Naturf. Freunde Berlin* **4**, 145–146.

Maynard-Smith, J. (1964) Group selection and kin selection. *Nature* **201**, 1145–1147.

Mayr, E. (1942) *Systematics and the Origin of Species.* Columbia University Press, New York.

Meester, J.A.J. (1960) The Col. Jack Scott Somalia Expedition. *Bull. Trans. Mus.* No. 4, 5–8.

Meester, J.A.J. and Hallett, A.F. (1970) Notes on early postnatal development in certain southern African Muridae and Cricetidae. *J. Mammal.* **51**, 703–711.

Michener, C.D. (1969) Comparative social behaviour of bees. *Annu. Rev. Entomol.* **14**, 299–342.

Michener, C.D. (1974) *The Social Behavior of the Bees.* Harvard University Press, Cambridge, Mass.

Moehlman, P.D. and Hofer, H. (1997) Cooperative breeding, reproductive suppression, and body mass in canids. In *Cooperative Breeding in Mammals.* (eds N.G. Solomon and J.A. French), pp. 76–128. Cambridge University Press, New York.

Moolman, M., Bennett, N.C. and Schoeman, A.S. (1998) The social structure and dominance hierarchy of the highveld mole-rat, *Cryptomys hottentotus pretoriae* (Rodentia: Bathyergidae). *J. Zool. Lond.* **246**, 193–201.

Moritz, C., Dowling, T.E. and Brown, W.M. (1987) Evolution of animal mitochondrial DNA: relevance for population biology and systematics. *Annu. Rev. Ecol. Syst.* **18**, 269–292.

Morrison, P.R. and Ryser, F.A. (1952) Weights and body temperatures in mammals. *Science* **116**, 231–232.

Mossman, H.W. and Duke, K.L. (1973) *Comparative Morphology of the Mammalian Ovary*. University of Wisconsin Press. USA.

Nanni, R.F. (1988) The interaction of mole-rats (*Georychus capensis* and *Cryptomys hottentotus*) in the Nottingham Road region of Natal. Unpubl. MSc thesis, University of Natal, Pietermaritzburg, South Africa.

Narins, P.M., Reichman, O.J., Jarvis, J.U.M. and Lewis, E.R. (1992) Seismic signal transmission between burrows of the Cape mole-rat, *Georychus capensis*. *J. Comp. Physiol.* **170**, 13–21.

Nedbal, M.A., Allard, M.W. and Honeycutt, R.L. (1994) Molecular systematics of hystricognath rodents: evidence from the mitochondrial 12S rRNA gene. *Mol. Phylogenet. Evol.* **3**, 206–220.

Nevo, E. (1961) Mole-rat *Spalax ehrenbergi*: mating behaviour and its evolutionary significance. *Science* **163**, 484–486.

Nevo, E. (1979) Adaptive convergence and divergence of subterranean mammals. *Annu. Rev. Ecol. Syst.* **10**, 269–308.

Nevo, E. (1991) Evolutionary theory and processes of active speciation and adaptive radiation in subterranean mole-rats, *Spalax ehrenbergi* superspecies in Israel. *Evol. Biol.* **25**, 1–125.

Nevo, E., Ben-Shlomo, R., Beiles, A., Jarvis, J.U.M. and Hickman, G.C. (1987) Allozyme differentiation and systematics of the endemic subterranean mole-rats of South Africa (Rodentia: Bathyergidae). *Z. Säugetierk.* **51**, 36–49.

Nevo, E., Filippucci, M.G. and Beiles, A. (1990a) Genetic diversity and its ecological correlates in nature: comparisons between subterranean, fossorial and above ground small animals. In *Evolution of Subterranean Mammals at the Organismal and Molecular Levels* (eds E. Nevo and O.A. Reig), pp. 347–366. Wiley Liss, New York.

Nevo, E., Honeycutt, R.L., Yonekawa, H., Nelson, K. and Hanzawa, N. (1990b) Mitochondrial DNA polymorphisms in subterranean mole-rats of the *Spalax ehrenbergi* superspecies in Israel and its peripheral isolates. *Mol. Biol. Evol.* **10**, 590–604.

Nevo, E., Honeycutt, R.L., Yonekawa, H., Nelson, K. and Hanzawa, N. (1993) Mitochondrial DNA polymorphisms in subterranean mole-rats of the *Spalax ehrenbergi* superspecies in Israel and its peripheral isolates. *Mol. Biol. Evol.* **10**, 590–604.

Nevo, E., Filippucci, M.G., Redi, C., Simson, S., Heth, G. and Beiles, A. (1995) Karyotype and genetic evolution in speciation of subterranean mole-rats of the genus *Spalax* in Turkey. *Biol. J. Linn. Soc.* **54**, 203–229.

Ogilby (1838) *Proc. Zool. Soc. Lond.* (6) pp. 5.

O'Riain, M.J. (1996) Pup ontogeny and factors influencing behavioural and morphological variation in naked mole-rats, *Heterocphalus glaber* (Rodentia, Bathyergidae). Uupubl. PhD thesis, University of Cape Town, South Africa.

O'Riain, M.J. and Jarvis, J.U.M. (1997) Colony member recognition and xenophobia in the naked mole-rat. *Anim. Behav.* **53**, 487–498.

O'Riain, M.J. and Jarvis, J.U.M. (1998) The dynamics of growth in naked mole-rats: the effects of litter order and changes in social structure. *J. Zool. Lond.* **246**, 49-60.

O'Riain, M.J., Jarvis, J.U.M. and Faulkes, C.G. (1996) A dispersive morph in the naked mole-rat. *Nature* **380**, 619–621.

Orians, G.H. and Pearson, N.E. (1979) On the theory of central place foraging. In *Analysis of Ecological Systems* (eds D.H. Horn, R. Mitchell and G.R. Stairs), pp. 155–177. Ohio State University Press, Columbus.

Page, R.E. and Metcalf, R.A. (1984) A population investment sex ratio for the honey bee (*Apis mellifera* L.). *Am. Nat.* **124**, 68–702.

Pamilo, P. and Crozier, R.H. (1996) Reproductive skew simplified. *Oikos* **75**, 533–535.

Pallas (1778) *Novae Species Quadrupedum e Glirum Ordine.* Erlangen: Wolfgang Walther, Part 2, pp. 172.

Parona, C. and Cattaneo, G. (1893) Note sull *Heterocephalus. Annali Dei Museo Civico Di Storia Naturale Di Genova* **13**.

Pearson, O.P. (1959) Biology of subterranean rodents, *Ctenomys*, in Peru. *Mem. Mus Hist. Nat. Javier Prado* **9**, 1–56.

Pearson, O.P. and Christie, M.I. (1985) Los tuco-tucos (genero-*Ctenomys*) de los Parques Nacionales Lanin y Nahuel, Argentina. *Hist. Nat. Corrientes.* **5**, 337–343.

Peeters, C. and Tsuji, K. (1993) Reproductive conflict among ant workers in Diacamma sp from Japan – dominance and oviposition in the absence of the gamergate. *Insectes Sociaux* **40**, 119–136.

Pepper J.W., Braude, S.H., Lacey, E.A. and Sherman, P.W. (1991) Vocalizations of the naked mole-rat. In *The Biology of the Naked Mole-Rat* (eds P.W. Sherman, J.U.M. Jarvis and R.D. Alexander), pp. 243–274. Princeton University Press, Princeton, NJ.

Perrin, M.R. and Curtis, B.A. (1980) Comparative morphology of the digestive system of 19 species of southern African myomorph rodents in relation to diet and evolution. *S. Afr. J. Zool.* **15**, 22–33.

Peters (1881) Sitz. Ber. Ges. Naturf. Fr. Berlin, pp. 133.

Peters, W.H.C. (1852) Naturwissenschafteliche Reise nach Mossambique. *Zoologie I. Saugetiere.* Georg Reimer, Berlin.

Pfennig, D.W. and Sherman, P.W. (1995) Kin recognition. *Scientific American*, June, 68–73.

Pitcher, T. and Buffenstein, R. (1994) The effect of dietary calcium content and oral vitamin D_3 supplementation on mineral homeostatis in a subterranean mole-rat *Cryptomys damarensis. Bone Miner.* **27**, 145–157.

Pitcher, T. and Buffenstein, R. (1995) Intestinal calcium transport in mole-rats (*Cryptomys damarensis* and *Heterocephalus glaber*) is independent of both genomic and non-genomic vitamin D mediation. *Exp. Physiol.* **80**, 597–608.

Platt, W.J. (1975) The colonization and formation of equilibrium plant species associations on badger mounds in a tall grass prairie. *Ecol. Monogr.* **45**, 285–305.

Potts, W.K., Manning, C.J. and Wakeland, E.K. (1991) Mating patterns in seminatural populations of mice influenced by MHC genotype. *Nature* **352**, 619–621.

Powell, R.A. and Fried, J.J. (1992) Helping by juvenile pine voles (*Microtus pinetorum*), growth and survival of younger siblings, and the evolution of pine vole sociality. *Behav. Ecol.* **3**, 325–333.

Purvis, A. and Rambaut, A. (1995) Comparative analysis by independent contrasts (CAIC): an Apple Macintosh application for analysing comparative data. *Computer Appl. Biosciences* **11**, 247–251.

Pyke, G.H. (1984) Optimal foraging theory: a critical review. *Ann. Rev. Ecol. Syst.* **15**, 523–575.

Pyke, G.H., Pulliam, H.R. and Charnov, E.L. (1977) Optimal foraging: a selective review of theory and tests. *Quart. Rev. Biol.* **52**, 137–154.

Rado, R., Levi, N., Hauser, H., Witcher, J., Alder, N., Intrator, N., Wollberg, Z. and Terkel, J. (1987) Seismic signalling as a means of communication in a subterranean mammal. *Anim. Behav.* **35**, 1249–1266.

Rado, R., Himelfarb, M., Arensburg, B., Terkel, J. and Wollberg, Z. (1989) Are seismic communication signals transmitted by bone conduction in the blind mole-rat? *Hear. Res.* **41**, 23–30.

Rahm, U. (1969) Gestation period and litter size of the mole-rat, *Tachyoryctes ruandae*. *J. Mammal.* **50**, 383–384.

Rasa, O.A.E. (1973) Intra-familial sexual reppression in the dwarf mongoose, *Helogale parvula*. *Naturwissenschaften* **60**, 303–304.

Rasa, O.A.E. (1977) The ethology and sociobiology of the dwarf mongoose (*Helogale undulata parvula*). *Z. Tierpsychol.* **43**, 337–406.

Rasa, O.A.E. (1986) Coordinated vigilance in dwarf mongoose family groups: the 'watchmans song' hypothesis and the costs of guarding. *Z. Tierpsychol.* **71**, 340–344.

Redi, C.A., Garagna, S., Heth, G. and Nevo, E. (1986) Descriptive kinetics of spermatogenesis in four chromosomal species of the *Spalax ehrenbergi* superspecies in Israel. *J. Exp. Zool.* **238**, 81–88.

Reeve, H.K. (1992) Queen activation of lazy workers in colonies of the eusocial naked mole-rat. *Nature* **358**, 147–149.

Reeve, H. K., Westneat, D.F., Noon, W.A., Sherman, P.W. and Aquadro, C.F. (1990) DNA 'fingerprinting' reveals high levels of inbreeding in colonies of the eusocial naked mole-rat. *Proc. Natl. Acad. Sci. USA* **87**, 2496–2500.

Reeve, H.K. and Sherman, P.W. (1991) Intracolonial aggression and nepotism by the breeding female naked mole-rat. In *The Biology of the Naked Mole-rat* (eds P.W. Sherman, J.U.M. Jarvis and R.D. Alexander), pp. 337–357. Princeton University Press, Princeton, NJ.

Reeve, H.K. and Ratnieks, F.L.W. (1993) Queen–queen conflicts in polygynous societies: mutual tolerance and reproductive skew, In *Queen Number and Sociality in Insects* (ed. L. Keller), pp. 45–85. Oxford University Press, Oxford.

Reichman, O.J. (1988) A comparison of the effects of crowding and pocket gopher disturbance on mortality, growth, and seed production of *Berteroa incana*. *Amer. Midland Nat.* **120**, 58–69.

Reichman, O.J. and Smith, S. (1985) Impact of pocket gopher burrows on overlying vegetation. *J. Mammal.* **66**, 720–725.

Reichman, O.J. and Jarvis, J.U.M. (1989) The influence of three sympatric species of fossorial mole-rats (Bathyergidae) on vegetation. *J. Mammal.* **70**, 763–771.

Reichman, O.J. , Whitham, T.G. and Ruffner, G. (1982) Adaptive geometry of burrow spacing in two pocket gopher populations. *Ecology* **63**, 687–695.

Reig, O.A. (1970). Ecological notes on the fossorial octodontid rodent *Spalacopus cyanus* (Molina). *J. Mammal.* **51**, 592–601.

Reig, O.A., Busch, C., Ortelis, M.O. and Contreras, J.R. (1990) An overview of evolution, systematic biology, cytogenetics, molecular biology and speciation in *Ctenomys*. *Prog. Clin. Biol. Res.* **335**, 71–96.

Rickard, C.A. and Bennett, N.C. (1997) Recrudescence of sexual activity in a reproductively quiescent colony of the Damaraland mole-rat, by the introduction of a genetically unrelated male – a case of incest avoidance in 'queenless' colonies. *J. Zool. Lond.* **241**, 185–202.

Roberts, A. (1951) *The Mammals of South Africa*, pp. 700. Trustees of the 'Mammals of South Africa' book fund. Central News Agency, Cape Town.

Rood, J. P. (1974) Banded mongoose males guard young. *Nature* **248**, 176.

Rood, J. P. (1978) Dwarf mongoose helpers at the den. *Z. Teirpsychol.* **48**, 277–287.

Rood, J.P. (1986) Ecology and social evolution in the mongooses. In *Ecological Aspects of Social Evolution* (eds D.I. Rubenstein and R.W. Wrangham), pp. 131–152. Princeton University Press, Princeton, NJ.

Rosenthal, C.M., Bennett, N.C. and Jarvis, J.U.M. (1992) The change in dominance hierarchy over time of a complete field-captured colony of *Cryptomys hottentotus hottentotus. J. Zool. Lond.* **228**, 205–225.

Rosevear, D.R. (1969) *The Rodents of West Africa.* Trustees of the British Museum (Natural History), London.

Rothe, H. (1971) Some remarks on the spontaneous use of the hand in the common marmoset (*Callithrix jacchus*). In *Proceedings of the Third International Congress of Primatology Zurich,* Vol. 3, pp. 136–141. Karger, Basel.

Royama, T. (1970) Factors governing the hunting behaviour and selection of food by the great tit (*Parus major* L.). *J. Anim. Ecol.* **39**, 619–668.

Rüppell, E. (1842) *Heterocephalus* nov. gen. uber saugethiere aus der ordnung der nager (1834). *Museum Senkenbergianum Abhandlungen* No. 3, pp. 99.

Sage, R.D., Contreras, J.R., Roig, V.G. and Patton, J.L. (1986) Genetic variation in South American burrowing rodents of the genus *Ctenomys* (Rodentia: Ctenomyidae). *Z. Säugetierk.* **51**, 158–172.

Sandow, J. (1983) The regulation of LHRH action at the pituitary and gonadal receptor level: a review. *Psychoneuroendocrinology* **8**, 277–287.

Schieffelin, J.S. and Sherman, P.W. (1995) Tugging contests reveal feeding hierarchies in naked mole-rat colonies. *Anim. Behav.* **49**, 537–541.

Schjelderup-Ebbe, T. (1922) Beitrage zur sozialpsychologie des Haushuhns. *Z. Psychol.* **88**, 225–252.

Schmidt-Nielsen, K. (1975) Desert rodents: physiological problems of desert life. In *Rodents in Desert Environments* (eds I. Prakash and P.K. Gosh), pp. 379–388. Junk, The Hague.

Schmidt-Nielsen, K., Hainsworth, F.R. and Murrish, D.E. (1970). Counter-current heat exchange in the respiratory passages: effects on water and heat balance. *Resp. Physiol.* **9**, 263–276.

Schramm, P. (1961) Copulation and gestation in the pocket gopher. *J. Mammal.* **42**, 167–170.

Schreber, J.C.D. von (1782) *Die Saugethiere,* Part 4. Erlangen. Wolfgang Walther 1792, 593–936.

Shanas, U., Heth, G., Nevo, E., Shalgi, R. and Terkel, J. (1995) Reproductive behaviour in the female blind mole (*Spalax ehrenbergi*) *J. Zool. Lond.* **237**, 195–210.

Sherman, P.W., Jarvis, J.U.M. and Alexander, R.D. (1991) Preface. In *The Biology of the Naked Mole-Rat* (eds P.W. Sherman, J.U.M. Jarvis and R.D. Alexander). Princeton University Press, Princeton.

Sherman, P.W., Jarvis, J.U.M. and Braude, S.H. (1992) Naked mole-rats. *Scientific American,* August, pp. 42–48.

Sherman, P.W., Lacey, E.A., Reeve, H.K. and Keller, L. (1995) The eusociality continuum. *Behav. Ecol.* **6**, 102–108.

Shortridge, G.C. (1934) *The mammals of South West Africa.* W. Heinemann, London.

Simson, S., Lavie, B. and Nevo, E. (1993) Penial differentiation in speciation of subterranean mole-rats, *Spalax ehrenbergi* in Israel. *J. Zool. Lond.* **229,** 493–503.

Skinner, J.D. and Smithers, R.H.N. (1991) *The Mammals of the Southern African Subregion.* University of Pretoria, Pretoria, South Africa.

Smith, M.S. (1984) Effects of the intensity of the suckling stimulus and ovarian steroids on pituitary GnRH receptors during lactation. *Biol. Reprod.* **31**, 548–555.

Smith, T.E., Faulkes, C.G. and Abbott, D.H. (1997) Combined olfactory contact with the parent colony and direct contact with non-breeding animals does not maintain suppression of ovulation in female naked mole-rats. *Hormones and Behavior* **31**, 277–288.

Smolen, M.J., Genoways, H.H. and Baker, R.J. (1980) Demographic and reproductive parameters of the yellow-cheeked pocket gopher (*Pappogeomys castanops*). *J. Mammal.* **61**, 224–236.

Snowdon, C.T. (1996) Infant care in cooperatively breeding species. *Adv. Stud. Behav.* **25**, 643–689.

Solomon, N.G. and Getz, L.L. (1997) Examination of alternative hypotheses for cooperative breeding in rodents. In *Cooperative Breeding in Mammals* (eds N.G. Solomon and J.A. French), pp. 199–230. Cambridge University Press, New York.

Solomon, N.G. and French, J.A. (1997) The study of mammalian cooperative breeding. In *Cooperative Breeding in Mammals* (eds N.G. Solomon and J.A. French), pp. 1–10. Cambridge University Press, New York.

Spencer, S.R., Cameron, B.D., Eschelman, L., Cooper, C. and Williams, L.R. (1985). Influence of pocket gopher mounds on a Texas coastal prairie. *Oecologia* **60**, 111–115.

Spinks, A.C. (1998) Sociality in the common mole-rat, *Cryptomys hottentotus hottentotus*, Lesson 1826: the effects of aridity. Unpubl. PhD thesis, University of Cape Town, South Africa.

Spinks, A.C., Van der Horst, G. and Bennett, N.C. (1997) Influence of breeding season and reproductive status on male reproductive characteristics in the common mole-rat, *Cryptomys hottentotus hottentotus*. *J. Reprod. Fert.* **109**, 78–86.

Spinks, A.C., Bennett, N.C. and Jarvis, J.U.M. (1999a) Regulation of reproduction in female common mole-rats, *Cryptomys hottentotus hottentotus*; the effects of breeding season and reproductive status. *J. Zool. Lond.* (in press).

Spinks, A. C., Branch, T.A., Croeser, S., Bennett, N.C. and Jarvis, J.U.M. (1999b) Foraging in wild and captive colonies of the cooperatively breeding common mole-rat, *Cryptomys hottentotus hottentotus* (Rodentia: Bathyergidae). *J. Zool. Lond.* (in press).

Stacey, P.B. and Ligon, J.D. (1987) Territory quality and dispersal options in the acorn woodpecker, and a challenge to the habitat-saturation model of cooperative breeding. *Am. Nat.* **137**, 831–836.

Suzuki, H., Wakana, S., Yonekawa, H., Moriwaki, K., Sakurai, S. and Nevo, E. (1996) Variations in ribosomal DNA and Mitochondrial DNA among chromosomal species of subterranean mole-rats. *Mol. Biol. Evol.* **13**, 85–92.

Swofford, D.L. (1993) *PAUP: Phylogenetic Analysis Using Parsimony (Version 3.1)*. (Computer Program). Champaign: Illinois Natural History Survey.

Tardif, S.D. (1997) The bioenergetics of parental behavior and the evolution of alloparental care in marmosets and tamarins. In *Cooperative Breeding in Mammals* (eds N.G. Solomon and J.A. French), pp. 11–33. Cambridge University Press, New York.

Taylor, P.J., Jarvis, J.U.M. and Crowe, T.M. (1985) Age determination in the Cape mole-rat *Georychus capensis*. *S. Afr. J. Zool.* **20**, 261–267.

Thigpen, L.W. (1940) Histology of the skin of a normally hairless rodent. *J. Mammal.* **21**, 449–456.

Thomas, O. (1895) On African mole-rats of the Genera *Georychus* and *Myoscalops*. *Ann. Mag. Nat. Hist.* pp. 238–241.

Thomas, O. (1897) Exhibition of small mammals collected by Mr Alexander Whyte during his expedition to the Nyika plateau and Masuka Mountains, nr Nyasa. *Proc. Zool. Soc. Lond.* pp. 430–436.

Thomas, O. (1910) List of mammals from Mount Kilimanjaro, obtained by Mr. Robin Kemp, and presented to the British Museum by Mr. C.D. Rudd. *Ann. Mag. Nat. Hist.* **8**, 303–316.

Thomas, O. (1911) On mammals collected by the Rev. G.T. Fox in northern Nigeria. *Ann. Mag. Nat. Hist., Ser.* 8, **7**, 457–463.

Thomas, O. and Schwann, H. (1904) On mammals from British Namaqualand. *Proc. Zool. Soc. Lond.* pp. 171–183.

Trivers, R. (1971) The evolution of reciprocal altruism. *Quart. Rev. Biol.* **46**, 35–47.

Urison, N.T. and Buffenstein, R. (1994) Kidney concentrating ability of a subterranean xeric rodent, the naked mole-rat (*Heterocephalus glaber*). *J. Comp. Physiol. B* **163**, 676–681.

Van Aarde, R. (1987) Pre- and postnatal growth of the the Cape porcupine *Hystrix africaeaustralis*. *J. Zool. Lond.* **211**, 25–33.

Van Couvering, J.A.H. (1980) Community evolution in East Africa during the late Cenozoic. In *Fossils in the Making: Vertebrate Taphonomy and Palaeocecology* (eds A.K. Behrensmeyer and A.P. Hill), pp. 272–298. Chicago University Press, Chicago.

Van Couvering, J.A.H. and Van Couvering, J.A. (1976) Community evolution in East Africa during the late Cenozoic. In *Human Origins: Louis Leakey and the East African Evidence* (eds G.L. Isaac and E.R. McCown), pp. 155–207. W.A. Benjamin, Menlo Park, Calif.

Van der Horst, G. (1972) Seasonal effects of anatomy and histology on the reproductive tract of the male rodent mole. *Zool Afr.* **7**, 491–520.

Van der Westhuizen, L.A. (1997) Social suppression of reproduction in the naked mole-rat, *Heterocephalus glaber*: plasma LH concentrations and differential pituitary responsiveness to exogenous GnRH. Unpubl. Masters thesis, University of Cape Town, South Africa.

Vandenbergh, J.G. (1988) Pheromones and mammalian reproduction. In *The Physiology of Reproduction* (eds E. Knobil and J.A. Neill), pp. 1679–1695. Raven Press Ltd, New York.

Vaughan, T.A. and Hansen, R.M. (1961). Activity rhythm of the plains pocket gopher. *J. Mammal.* **42**, 541–543.

Vehrencamp, S.L. (1983) Optimal degree of skew in cooperative societies. *Am. Zool.* **23**, 327–335.

Vleck, D. (1979) The energy cost of burrowing by the pocket gopher *Thomomys bottae*. *Physiol Zool.* **52**, 122–125.

Vleck, D. (1981) Burrow structure and foraging costs in the fossorial rodent *Thomomys bottae*. *Oecologia* **49**, 391–396.

Wallace, E. and Bennett, N.C. (1998) The colony structure and social organization of the giant Zambian mole-rat, *Cryptomys mechowi. J. Zool. Lond.* **244**, 51–61.

Wasser, S.K. and Barash, D.P. (1983) Reproductive suppression among female mammals: Implications for biomedine and sexual selection theory. *Quart. Rev. Biol.* **58**, 513–538.

Watt, J.M., and Breyer-Brandwijk, M.G. (1962) *The medical and poisonous plants of southern and eastern Africa*, 2nd Edition. E. and S. Livingstone Ltd, Edinburgh.

Wcislo, W.T. and Danforth, B.N. (1997) Secondarily solitary: the evolutionary loss of social behavior. *Trends Ecol. Evol.* **12**, 468–474.

Weir, B.J. (1974) Reproductive characteristics of hystricomorph rodents. *Symp. Zool. Soc. Lond.* **34**, 265–301.

Weir, B.J. and Rowlands, I.W. (1974) Functional anatomy of the hystricomorph ovary. *Symp. Zool. Soc. Lond.* **34**, 303–332.

West Eberhard, M.J. (1975) The evolution of social behaviour by kin selection. *Quart. Rev. Biol.* **50**, 1–33.

Westlin, L., Bennett, N.C. and Jarvis, J.U.M. (1994) Relaxation of reproductive suppression in non-breeding naked mole-rats. *J. Zool. Lond.* **234**, 177–188.

Wildt, D.E., Bush, M., Goodrowe, K.L., Packer, C., Pusey, A.E., Brown, J.L., Joslin, P. and O'Brien, S.J. (1987) Reproductive and genetic consequences of founding isolated lion populations. *Nature* **329**, 328–331.

Williams, S.L., Schlitter, D.A. and Robbins, L.W. (1983). Morphological variation in a natural population of *Cryptomys* (Rodentia: Bathyergidae) from Cameroon. *Ann. Mus. Roy. Centr. Sc. Zool.* **237**, 159–172.

Wilson, E.O. (1971) *The Insect Societies*. Harvard University Press, Cambridge, Mass.

Withers, P.C. and Jarvis, J.U.M. (1980) The effect of huddling on thermoregulation and oxygen consumption for the naked mole-rat. *Comp. Biochem. Physiol.* **66**, 215–219.

Woolfenden, G.E. (1975) Florida scrub jays – helpers at the nest. *Auk* **92**, 1–15.

Wood, A.E. (1985) The relationships, origin, and dispersal of the hystricognathus rodents. In *Evolutionary Relationships among Rodents: A Multidisiplinary Analysis* (eds W.P. Luckett and J.-L. Hartenberger), pp. 475–513. NATO ASI Series. New York.

Woolsey, T.A., Welker, C. and Schwartz, R.H. (1975) Comparative anatomical studies of the SmI face cortex with special reference to the occurrence of 'barrels' in layer IV. *J. Comp. Neurol.* **164**, 79–94.

Wroughton, R.C. (1907) On a collection of mammals made by Mr. S.A. Neave in Rhodesia, north of the Zambesi, with field notes by the collector. *Manchester Memoirs* **51**, 1–39.

Yahav, S. and Buffenstein, R. (1991) Huddling behavior facilitates homeothermy in the naked mole rat *Heterocephalus glaber*. *Physiol. Zool.* **64**, 871–884.

Yahav, S. and Buffenstein, R. (1992) Caecal function provides the energy of fermentation without liberating heat in the poikilothermic mammal, *Heterocephalus glaber*. *J. Comp. Physiol. B* **162**, 216–218.

Zarrow, M.X. and Clark, J.H. (1968) Ovulation following vaginal stimulation in a spontaneous ovulator and its implications. *J. Endocrinol.* **40**, 343–352.

Zullinger, E.M., Ricklefs, R.E., Redford, K.H. and Mace, G.M. (1984) Fitting sigmoidal equations to mammalian growth curves. *J. Mammal.* **65**, 607–636.

Index

Page numbers in *italics* refer to figures and tables.